DO-IT-YOURSELF ENERGY-SAVING PROJECTS

BY THE EDITORS OF SUNSET BOOKS AND SUNSET MAGAZINE

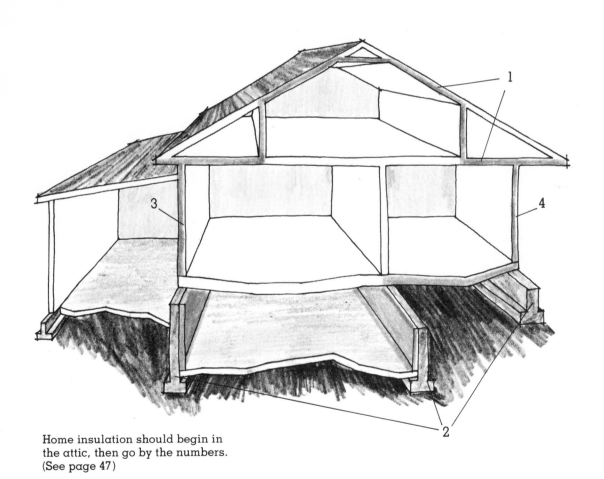

Home insulation should begin in the attic, then go by the numbers. (See page 47)

LANE PUBLISHING CO., MENLO PARK, CALIFORNIA

Supervising Editor: **Bob Thompson**

Research and Text: **Michael Scofield, Bill Tanler**

Design: **Russell Leong**

Illustrations: **Joe Seney, Terry Meagher**

Cover: Photograph by **Jack McDowell;**
   cover design by **Joe di Chiarro.**

For whom the book toils

If you live in an average American home—a conventional building on a conventional urban or suburban lot—there is good news and bad news in the energy efficiency department.

The bad news is that most buildings older than 5 years were not built with energy efficiency in mind.

The good news is that the average American home is not a hopeless case. Far from that, chances for major improvement in your building's performance are excellent. In fact, if your home really is average you can expect to trim energy use about 10 percent and do it for very little cash outlay. And if you're willing to invest time and money in building projects of some scope, possible energy savings range up to 30 percent.

This project-oriented book deals almost entirely in practical steps short of remodeling that you can take to trim your energy use.

• It is mainly for people who live in conventional houses, though it does not ignore those with opportunities presented by buildings of unusual design.

• It is plainly written for people who want to make profitable decisions without having to qualify as architectural engineers or physicists before beginning.

• As much as possible, it shows cost effectiveness of projects, though material and labor prices across the country are too variable to permit absolute statements.

• Finally, the book is for people who feel challenged to do some, perhaps all, of the work themselves in order to make the payback time as short as possible.

Acknowledgments

We owe thanks to many for their help in preparing this book, especially Holly Lyman Antolini; William Comstock, Association of Home Appliance Manufacturers; Richard Hutchinson, BDP Company; Milton Palmer and Raymond Sinclair, Baltimore Gas and Electric Company; Edgar J. Orme III, Better Homes; Paul Shea, Boston Edison Company; Brookhaven National Laboratories; Nancy Licht, California Energy Commission; Maury Callaghan; E. G. Edson, Detroit Edison; Marian Rees, Energy Information Center, Stanford University; Paul Engelhart, Engelhart Electric Company, Inc.; R. C. Bythewood, Georgia Power Company; Valerie Walsh, Green Horizon; Brooks Bishop and JoAnn Smith, Gulf States Utilities Company; Lawrence Berkeley Laboratories; Eugene E. Mallette; Walter Marks; Edward Medved; Alan Davies, National Bureau of Standards; Tom Auzenne, Lee Callaway, Eric A. Heim, James M. Loorya, and Michael Mertz, Pacific Gas and Electric Company; David Paulsen; Gerry Vurciaga, Public Service Company of Colorado; James Croft, Puget Sound Power & Light Company; J. K. Ramstetter; Randall Bettis, Redwood Plumbing Company, Inc.; Gene Kelley, Union Electric Company; John Ryan and James Smith, United States Department of Energy; and the state energy office of each of the 50 states.

Editor, Sunset Books: David E. Clark

First Printing May 1981

CONTENTS

Natural ventilation can take the load off air conditioners—or replace them. (See page 13)

BEFORE YOU BEGIN

We left few stones unturned in developing ways to help you get the most out of home appliances. But . . .

Heating and its opposite face, cooling, take most of the space in this book because (as the facing page clearly shows) they offer the best chances for major cuts in home energy use.

To make the most of your opportunities, look at your house as one big machine with three main working systems:

1. air exchanger
2. heat exchanger
3. heat (or cool) producer

In each system the goal is to win controlled flexibility: to be able to keep heat outside when the summer sun broils, and inside when wintry storms rage. Because of the way the three systems are interconnected, the most efficient attack on energy waste must work through the three in the order in which they are listed above.

That is, you should look for possible weak points, starting with air leaks, moving next to insulation (including at windows), and ending with your heaters and coolers. Unless you find an uninsulated attic—the biggest single source of heat gain or loss—you probably will want to make improvements in the same sequence.

Plugging cracks so your house exchanges air with the outside only as fast as you wish is the simplest, least expensive improvement you can make. More to the point, if you fail here, bales of insulation and a huge furnace will not help much. Your conditioned air will escape almost as fast as it can be manufactured.

Once the house is reasonably weathertight, the next step is to improve control over how much heat you gain or lose through windows and other glass. Only after these steps have failed should it be necessary to take the more costly step of insulating the main frame, let alone the most costly step of adding to or replacing your heating and cooling system(s). In fact if the first two steps are done well, the usual discovery is that your heaters and coolers are more powerful than they need to be.

Measuring your chances

The real question before the house is, what pays? Which steps, which projects are going to yield real dividends in a reasonable period.

Typical Annual Energy Bills

If you are a family of four living in a three-bedroom, two-bath home with about 1,500 square feet of heated living space, the total and detailed energy use figures for your climate region are likely to be close to your own annual energy consumption. If your house is bigger, the heating and cooling bill increases proportionately; the other figures change little if at all with changes in the size of a building.

Appliance		New England	Atlantic Seaboard	Great Lakes	Plains	Southeast	Gulf States	Rockies	Pacific Northwest	Southwest	
						Region					
Central heating	(In kilowatt hours)	21,100	17,731	15,125	16,900	15,000	10,500	10,000	10,340	5,450	
Air conditioning		1,080	2,596	1,500	3,000	4,800	10,200	1,000	0	2,089	
Water heating		5,700	6,000	6,065	6,741	6,000	4,000	5,000	6,323	4,595	
Range		1,175	1,380	1,068	1,200	1,440	1,000	750	980	679	
Refrigerator-freezer		1,500	1,900	1,428	1,800	1,815	1,830	1,800	1,134	1,200	
Dishwasher		363	324	180	300	375	432	100	432	337	
Clothes washer		103	125	108	72	84	100	100	245	170	
Clothes dryer		1,000	852	900	962	1,440	550	950	1,080	1,111	
Total		32,021	30,908	26,374	30,975	30,954	28,612	19,700	20,534	15,631	

1 therm of natural gas = 30 kilowatt hours; 1 gallon of heating oil = 42 kwh

Blanket statements are impossible for a number of reasons. The strength and weaknesses of individual houses are unknowns. Climate is vitally related to energy use patterns. Utility rates vary, and so do the costs of building materials.

In nearly every state, either the state energy office or your local utility is prepared to help you perform an energy audit on your house; in many states, one agency or the other will send a trained inspector to do the work. There is no better or more accurate way to find out which jobs most need doing around your house or apartment.

Meanwhile, this book is patterned on a compilation of several such audit forms. It offers general answers to the questions raised by audits, and it gives practical cures for fuel-wasting conditions commonly identified by these on-the-spot checks.

On the facing page is a table of average energy use figures for all residential customers of several utility companies around the nation. The locations were chosen to reflect major climate regions as measured by the amounts of heating and cooling required.

Below is a chart showing—in very general terms—some maximum potential savings.

To help you determine your own needs, we offer this chart and more detailed ones with each chapter as "worst case" examples of potential savings; the numbers are in units of energy as they appear on utility bills. Take into account any advantage you have over the worst case. (For example, a completely uninsulated attic is a worst case; any insulation you have counts as an advantage.) When you find your saving in terms of energy units, multiply the figure by local rates. Compare your final figure with the cost of the improvement to find out if it's worth doing.

Small jobs ought to pay for themselves within 3 years. Larger ones will still be profitable over a 7-year payback period.

The yardsticks

Making practical improvements in your home's energy efficiency depends on some understanding of the units of measure used by utilities to bill you.

• Therm. A unit of natural gas equal to 100 cubic feet. This much gas has a potential of 100,000 BTUs of heat. (BTUs, or British thermal units, are defined in the glossary.)

• Kilowatt hour. The electrical counterpart to therms is 1,000 watts of electricity transmitted for 1 hour. A kilowatt hour has a heat potential of 3,413 BTUs.

• Gallon. The billing unit for heating oil, which has a potential of 140,000 BTUs of heat.

So, 30 kilowatt hours roughly equal 1 therm of gas, while one gallon of oil equals 1.4 therms.

How Much Can You Save?

If you are buying about as much energy each year as the average amount shown for your climate region, your chances of saving are substantial. A diligent attic to basement effort to trim energy use to the minimum can save you as much as we show below, though some changes would be uneconomic. As quick guides to where and how you can save, see the tables on pages 6, 20, 46, 66, and 86.

Appliance		Region								
		New England	Atlantic Sea-board	Great Lakes	Plains	South-east	Gulf States	Rockies	Pacific North-west	South-west
Central heating	(In kilowatt hours)	8,440	7,092	6,050	6,760	6,000	4,200	4,000	4,136	2,180
Air conditioning		216	519	300	600	960	2,040	200	0	418
Water heating		855	900	910	1,011	900	600	750	948	689
Range		59	69	53	60	72	50	38	49	34
Refrigerator-freezer		375	475	357	450	454	458	450	284	300
Dishwasher		54	49	27	45	42	65	15	65	51
Clothes washer		21	25	22	14	17	15	15	37	34
Clothes dryer		150	128	135	144	216	149	143	110	167
Total		10,170	9,257	7,854	9,084	8,661	7,577	5,611	5,629	3,873

1 therm of natural gas = 30 kilowatt hours; 1 gallon of heating oil = 42 kwh

MAKING AIR BEHAVE

If ever a winter wind has attacked you in your own living room, you know the value of getting rid of drafts. On the other hand, the joys of throwing the whole house open to a balmy spring day do not need much selling.

These worst and best conditions of managing air circulation inside a house point up the need for flexible control: the ability to let the air replace fuel-using devices when possible and the ability to keep out extremes of weather, which can overwork your heating and cooling system(s).

Keeping air out requires caulking and weatherstripping—homely tasks, but ones that pay major dividends for minor expense. A flexible system should control the entry of air no matter what the season: hot summer winds are not as noticeable indoors as wintry

ones, but weathertight buildings can spare air conditioners substantially in hot-summer regions.

Letting agreeable air into living space requires windows or doors openable in patterns that produce good ventilation but not gusting winds. This is not just a sport for balmy spring days. Any time outdoor air ranges between 68° and 78°F (20° and 26°C), it may save heater or air conditioner running time if only you will let it inside.

Attics are another, perhaps even more important, place where good ventilation can save energy by taking a sizeable part of the load off cooling devices in summer, and help insulation remain at peak effectiveness during the heating season.

It is important to note here that, if you don't open windows

or provide some ventilation, you can do too good a job of keeping your house weathertight. The goal should be one complete exchange of inside air per hour (as opposed to three or more in a noticeably drafty building).

If the exchange rate falls below one per hour you risk indoor air pollution. Cigarette smoke is the most obvious cause, but not the only one: construction materials may contain formaldehyde or asbestos, and furnaces may give off carbon monoxide and nitrogen dioxide.

It is almost impossible to fall to an air exchange rate of less than one per hour unless all windows are double glazed and all doors have positive seals. But these conditions may be met in some new structures. If you suspect a problem, consult with your builder or architect.

Heating and Cooling Savings through Weathertightening

Percentages in the chart assume that the house is so poorly weatherstripped and caulked as to be drafty.

Action	Range of savings	Your potential savings
Caulking (See page 7)	Estimates on potential savings range from 4 to 7% of energy used on heating and cooling in a poorly insulated house, much more in well-insulated buildings.	
Weatherstripping (See page 8)	Potential savings are 4 to 7% in poorly insulated buildings, more in well-insulated ones.	
Attic ventilation to 1.5 cubic feet per minute Using static vents (See page 17)	If ceiling is insulated to R-30, savings are marginal or nonexistent. If attic is poorly insulated or uninsulated, savings can range upward from 16% of cooling bill.	
Using power vents (See page 17)	May result in marginal savings on air conditioner use; can replace air conditioner in some climates.	

Caulks

The number and variety of caulking compounds available to seal air leaks in your home has grown rapidly in recent years. You may have to take a little more care in selecting, but you can expect to find caulks formulated for highly specific tasks.

Most caulks fall into one of these basic categories: elastomers, butyl rubber, latex base, and oil base. Most are available in standard cartridges that fit an inexpensive trigger-operated caulking gun. One exception is polyurethane, which is dispensed from a pressure can.

The elastomers. This group includes silicones, polysulfide, and polyurethane. The elastomers share two basic characteristics: they are long lasting and they are the most expensive. However, the extra cost may be worth it, for they last 20 years or more if applied properly. The silicones are highly elastic and are the most common elastomers. They can't be painted, but do come in clear, white, and a few colors.

Polyurethane is suited for filling large cracks, because it expands as it dries; it must be protected from the sun if used outdoors, though. Polysulfide is not recommended for the novice, as it is difficult to work with and toxic until cured.

Butyl rubber. This middle-priced caulk should last up to 10 years. Though less flexible than silicone, it can be painted and is effective on masonry or metal surfaces.

Latex base. Two readily available varieties are acrylic latex and plain latex. The acrylics are slightly longer lasting and more expensive. Both should be painted if exposed to weather. Latex caulks, unlike all others, can be applied to damp surfaces.

Oil base. This is the cheapest caulk and the least desirable because it shrinks and lasts only 1 or 2 years. Its low price makes it a good stopgap solution to a temporary problem, though.

Whatever caulks you choose, examine the container for specific instructions and clean-up directions. Protect unused caulk in tubes by inserting a nail in the nozzle and then capping with plastic film or aluminum foil.

For caulking electrical outlets, use outlet insulators—special die-cut pieces of pliable foam that fit behind face plates. These are also available for switch plates.

Where to caulk

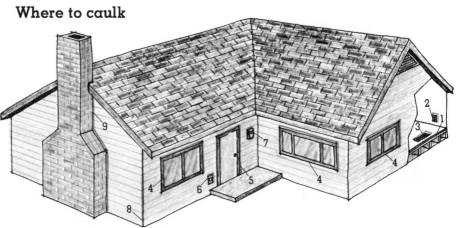

Every joint between two materials needs caulk. One study shows that, in frame construction, 25 percent of air infiltration is at sole plate (1), 20 percent around electrical outlets (2), 13 percent at heater duct system (3), and 12.5 percent at exterior windows (4). Other places to caulk are exterior doors (5), vents or other devices that penetrate walls (6), outside light fixtures (7), exterior corners (8), and joints between fireplace chimney and wall (9).

How to caulk

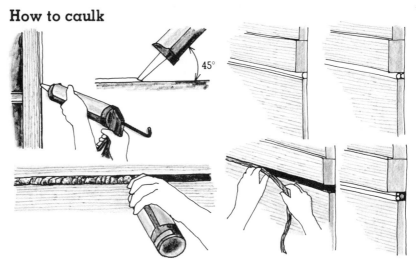

If using a caulking gun, cut nozzle diameter to equal width of gap to be filled. Hold gun or spray can at about a 45° angle. Work slowly so nozzle pushes caulk well into gap.

In deep cracks, apply one bead (strip of caulking) well inside crack; then apply a second bead at surface. In wide cracks, stuff with caulking rope or oakum.

Weatherstripping

Lumped together under the general name of weatherstripping is a broad variety of materials designed to plug gaps between doors or windows and their frames.

Most weatherstripping can be selected on the basis of budget, durability, esthetics, and ease of installation. Some types are limited to specific uses; for example, to close gaps beneath doors. Almost any weatherstripping you add will be a significant improvement over none at all.

Preparation also is important. Minor carpentry on doors or windows may be necessary to obtain a better fit within the frame, and thus a proper seal for the weatherstripping (see page 11).

Considering the low cost of most weatherstripping materials, it makes good sense to choose the best quality. However, if pennies count, approach each door or window separately.

For example, the less expensive foam tapes or felts may be adequate around rarely used windows. But for windows used almost daily, and for all doors, the best quality materials will provide the best insulation and will wear the longest.

The types of weatherstripping illustrated here are readily available and easy to install. There are, in addition, a few other effective styles that require above-average woodworking skills or a professional installer. Among them are several interlocking designs in which one metal strip fastens to the door's edge and another to the jamb, so that the two fit tightly together when the door is closed. Some of these require the cutting of grooves (in the door, the jamb, or both) that are extremely accurate and call for sophisticated power tools.

What's available in weatherstripping

Metal:

Spring metal weatherstripping is usually sold in rolls and made of brass, aluminum, or stainless steel. It is nailed in place, and a simple crimp in the metal makes a seal with the closed door or window. Spring metal poses two problems: once installed it cannot be adjusted easily, and if crimped or torn in use, it seldom can be repaired.

Cushion metal comes in rigid, V-shaped strips and is similar in use to spring metal. Some strips attach with nails; many have screw slots that allow adjustment while in place. The flange can be pried open wider to obtain a closer fit, but neither this nor spring metal will seal irregular gaps. Both have the advantage of being hidden, and both are effective in sash or horizontal sliding windows. Special spring-metal strips are available for metal-frame windows, usually from the window manufacturer.

Rigid gasket:

Wood-and-foam combines a strip of polyurethane foam bonded to a length of wood molding. It attaches with screws or nails and produces a reasonably tight seal. Though visible, it can have a quality look and can be painted. (Paint the wood only; not the foam.)

Wood-and-vinyl consists of a pliable vinyl strip set into a length of wood molding. It is similar to wood-and-foam weatherstripping. Both are usually attached with nails and cannot be moved easily once in place. Both are most commonly used around doors: they seal by compressing when the door (or window) closes.

Metal-and-vinyl weatherstripping is similar to wood-and-vinyl in that it utilizes a flexible vinyl strip inset in rigid molding. An advantage of metal is that it needn't be painted. Designs with slotted screw holes can be adjusted to compensate for wear.

Extruded vinyl differs from other rigid-gasket weatherstripping only in that it is made entirely from vinyl and is one piece. It attaches and functions in the same manner as the other rigid-gasket types. Slotted screw holes make it adjustable, and painting is optional; avoid getting paint on the surface that comes in contact with the door.

Pliable gasket:

Foam tape is one of the least expensive weatherstripping materials for use around doors or windows. Foam tape comes in rolls of varying thicknesses and widths, and usually attaches with self-adhesive or glue. It is among the least durable, especially if used on exterior doors. However, it can be inexpensive weatherstripping for use around seldom-opened windows.

Standard felt weatherstripping can be made of wool or a lower-priced combination of hair and cotton. It can be installed easily with glue, nails, or staples. Its main disadvantages are that it doesn't wear well and tends to mildew or rot in a damp climate.

Vinyl-covered foam is one of the more popular weatherstripping materials because it is versatile and relatively inexpensive. Used around windows as well as doors, it resists friction well and can form a good seal on uneven surfaces. As it usually is nailed or attached with self-adhesive, it can't be adjusted readily.

Felt clinched in aluminum suffers from some of the same shortcomings as standard felt weatherstripping, but lasts considerably longer because of its metal reinforcement. It still can be damaged by moisture. When installing, take care not to crimp the aluminum as this makes a less effective seal.

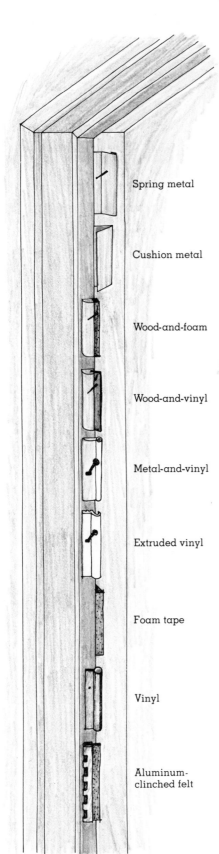

Spring metal

Cushion metal

Wood-and-foam

Wood-and-vinyl

Metal-and-vinyl

Extruded vinyl

Foam tape

Vinyl

Aluminum-clinched felt

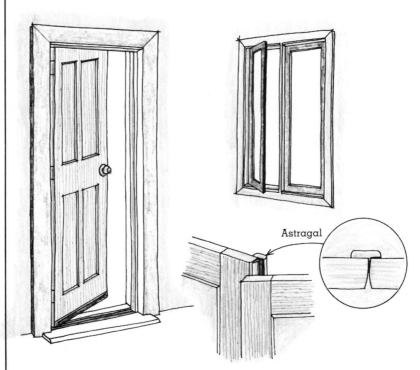

Astragal

Weatherstripping doors and hinged windows

Conventional weatherstripping is applied to all four sides of hinge-mounted windows, whether casement or hinged at top or bottom. It is applied only to the tops and sides of doors (see page 10 for threshold weatherstrips). On double doors an extra strip is applied to the inside face of the molding (astragal) that covers the crack between the two.

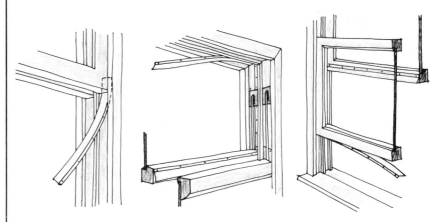

Weatherstripping sliding windows

Sliding windows typically call for spring or cushion metal weatherstripping secured all around the sash. In vertically mounted units, secure an extra strip across the outward face of the lower sash's top rail. This seals the gap between it and the upper sash. In horizontal windows, the extra strip fastens to the inner unit, but is still hidden from view on the outer face.

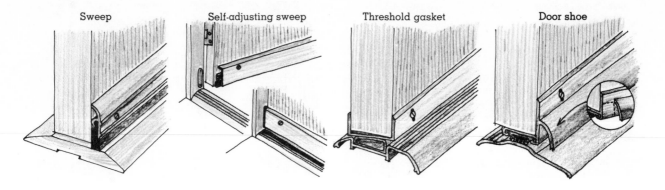

Sweep   Self-adjusting sweep   Threshold gasket   Door shoe

Thresholds need special attention

A number of weatherstripping systems, produced by several different manufacturers, are designed to seal the gap under exterior doors. They fall roughly into three types: door sweeps, vinyl gaskets, and door shoes.

Door sweeps. The least expensive and easiest to install, the door sweep can be cut to size and attached without removing the door from its hinges. Also available is a more sophisticated door sweep that raises when the door is opened. As most exterior doors open inward, the adjustable door sweep is designed to clear the threshold and carpeting, yet form a tight seal when the door is closed.

Vinyl gasket. The seal here is obtained by the gasket pressing up against the bottom of the door. It is installed permanently, either built into a metal threshold or secured in its metal frame to an existing wooden threshold. Though subject to wear from normal foot traffic, the vinyl gasket can be replaced easily. It is not usually necessary to remove the door in order to install a vinyl gasket, but it may be necessary to trim the bottom of the door to obtain the right clearance. Bevel the bottom of the door about ⅛ inch, tapering down toward the floor in the direction the door opens.

Door shoe. This unit attaches to the bottom of the door and seals with the threshold, using a gasket similar to the vinyl gasket installed directly on top of the threshold. There are door shoe systems that can be attached without removing the door, but usually the bottom of the door must be trimmed to assure a good seal.

Some door shoe systems include a built-in rain drip shield designed to keep water from seeping in under the door. If you install a vinyl gasket threshold or buy a door shoe without a rain drip, the shield is inexpensive and easy to install as a separate operation.

Installation tips. Properly installed, most systems do what they were designed to do. Still, the most effective tend to be the most expensive and require the greatest care in installation. Your final choice will be governed by budget, esthetic preferences, and compatibility with your particular style of door and threshold.

In all cases, the manufacturer provides the installation directions you'll need. Units that attach to the flat exterior surface of a door or directly to an existing wooden threshold require only basic tools. Replacing a threshold or installing hidden systems may require more woodworking skills. Sometimes cutting a groove in (or mortising) the bottom of the door is necessary, and elaborate tools (a router, for example) may be needed.

Make a weather worm

If you have a gap under a door that is too great for conventional weatherstripping methods, you may want to make your own "weatherstripping worm." All you need is a scrap of cloth 10 inches wide and a few inches longer than the width of the problem door. Simply sew the cloth into a tube with one end open, fill it with sand, and sew the last end closed. Place it along the threshold, and it will do a good job of cutting down on drafts. It can also be used under an interior door to prevent warm air from escaping into a cooler, unused room.

When worm is not snugged against inner side of door, it can hang on out of way on knob.

Preparing a door for weatherstripping

Weatherstripping around doors will be most effective if you make certain each door sits squarely in its frame. This is particularly important if you select spring metal or cushion metal weatherstripping that requires perfect clearance between the door and its jamb.

If the door and jamb are square, it should be possible to align them perfectly by securing loose door hinges or adjusting poorly set ones. Only if the door has swollen or become warped, should it be necessary to shape it by sanding or using a plane. If that's the case, avoid planing or sanding the latch side of the door, as it is usually beveled to assure a tight fit.

A belt sander is best for this job; coarse sandpaper followed by finer sandpaper can be used effectively if a belt sander isn't available. Wrap the sandpaper around a block to help keep the sanded surface even and square to the door. Don't be too hasty to plane; sanding is easier to control. If you must plane, remember these basic rules:

• When working on the top or bottom of a door, plane toward the center to avoid splitting at the corners;

• When working along a side, plane with the grain;

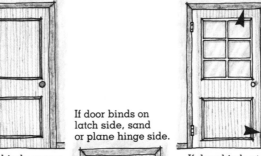

If door binds across top or bottom, sand or plane.

If door binds on latch side, sand or plane hinge side.

If door binds at arrows, shim top hinge or deepen bottom mortise.

If door binds at arrows, deepen top mortise or shim bottom hinge.

• On all surfaces, keep the body of the plane aligned with the surface you are cutting, and resting flat upon it.

In all cases it's better to correct fit problems by working on the door, not the door frame.

These techniques also work with hinged windows.

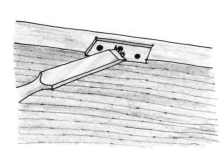

Resetting hinges

To cut a deeper notch, or mortise, for hinge to fit into, use a wood chisel to shave bottom of mortise. Work on door-side hinge leaf. Leaf should be left flush with edge of door or slightly recessed.

Several materials can be used as shims to lessen the depth of mortise. Sheet brass, which comes in varying thicknesses, is most durable. It is available in small squares at some hardware stores or metal supply shops. Shims can also be

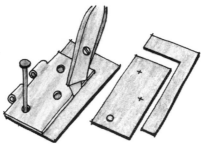

cut from any noncompressible cardboard, particle board, or similar material.

To cut a shim, use hinge leaf as a template and make shim slightly smaller in each dimension.

When putting hinge on again, use a drill or a nail to start screw holes in shim. Take care to center them in holes of hinge leaf.

It's unwise to glue shims in place, for they may need replacing again or you may wish to replace entire door at some later date.

Repairing worn screw holes

Inserting wooden match sticks in worn screw holes provides a bite for replacement screws' threads. Use screws slightly longer than originals. Strongest, most durable repair, especially for heaviest doors, is to carve a piece of wood or dowel to fit the old hole, coat it with woodworking glue, and push or hammer it into place. After glue dries, trim peg flush, drill a new guide hole, and replace hinge.

Tightening sash windows

The rattle you hear when the wind blows against your sash window is the only signal you need to know that the window is hung too loosely in its track. Inside air is escaping or outside air is sneaking in, even if there is weatherstripping in place.

The cure requires adjusting or reseating the stops.

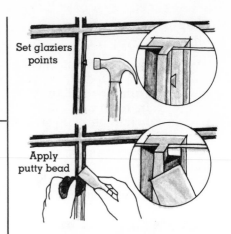

Set glaziers points

Apply putty bead

Adjusting stops

If gap is not too wide, cure is relatively simple. Take a hammer, a block of wood for a hammering surface, and some 6d finishing nails. Then cut a cardboard shim (the back from an old writing tablet works perfectly).

Place shim between stop molding and sash. Place block of wood against outer surface of stop molding and hit block with enough force to break paint between window frame and stop. Go from top to bottom of frame until stop rests snugly against shim. If stop shows a tendency to spring back to its original position, use three or four finishing nails to secure it in desired position. Repeat entire process on opposite side.

When you're finished, countersink and putty the new nails and touch up paint.

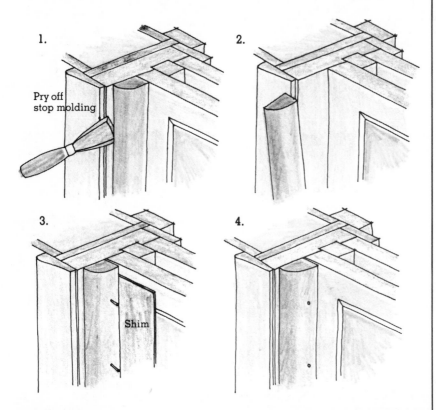

1. Pry off stop molding

2.

3. Shim

4.

Reseating stops

With a wide-bladed chisel, begin at bottom or top to pry stop molding away from window frame. To minimize paint chipping, score paint first where stop joins frame. Also be careful not to mar surface of frame when prying with chisel. Unless stop has become brittle with age, it should come away undamaged and in one piece.

Using a chisel or sandpaper block, clean any paint or ridges off frame on previously unexposed side of stop molding; then reseat molding on frame, using a cardboard shim for position. Using 6d finishing nails, nail the stop in place; then countersink and putty nails. Repaint stop molding to cover crack between stop and window frame.

Reputtying windows

Not all air infiltration is around the window and its sash. Sun, moisture, heat, and freezing combine over time to deteriorate the putty that holds window glass in place, leaving air gaps. The solution is reputtying.

There are several ready-to-use glazing compounds available today in cans. They help keep reglazing a simple job for metal-framed windows that require putty as well as for standard wood-framed windows. Metal windows that use neoprene or other continuous gaskets also may develop air leaks, but repair is a job for a professional glazier.

Working from top of frame and down along sides, take a wood chisel and chip away old putty. Replace any glazier's points (small metal triangles) that may have come loose in the process.

Coat exposed wooden frame lightly with linseed oil or wood sealer; then apply a sealing bead of putty. Press firmly for a good seal, using putty knife to smooth the finish. Putty should be flush with outer edge of molding and cover edge of glass in frame.

Putty will be dry enough to paint in a week to 10 days. Paint should reach 1/16 inch up glass to assure a good weather seal. After paint dries, smudges can be removed with a razor blade.

Venting living space

Circulating fresh air through your home is necessary for any of several logical reasons—to remove odors, clear out smoke, or simply to renew the oxygen supply. When we enjoyed the luxury of cheap energy and were less energy-conscious than we are today, poorly insulated and nonweathertight homes took care of the ventilation problem without much help from the homeowner.

With new homes being built to stricter energy-saving standards and with many homeowners caulking and weatherstripping the leaky gaps in their older homes, special attention may be needed to push fresh air through the house to keep a healthy interior environment. The purpose of making a home weathertight is not to prevent fresh air from circulating, but to control the flow of air.

Ventilation can be controlled with sophisticated equipment like air conditioners, heat pumps, air exchangers, and whole-house fans with thermostats. Or ventilation can be controlled at little or no cost—you simply open windows or utilize less complex, manually controlled window fans.

Natural venting

When outside temperatures are agreeable, it's possible to ventilate a home and produce a comfortable interior with absolutely no energy expenditure. Natural venting is no more than opening doors and windows—not all the doors and windows, but selected ones—to produce draft patterns that keep the air moving as you like it.

As the sketches show, there are three basic approaches: cross drafts, chimney venting, and single wall ventilation.

Chimney ventilation. The idea is simple: warm air rises. By whatever avenues possible, bring cool air into house at lowest possible level, and establish upward paths so warm air can escape at highest point. Upward air paths can follow stairwells to a dormer window, or can flow through attic trapdoor to a gable vent.

Ideal cross ventilation. To bring in air, open low windows (or window sections) on side of house facing wind; to let air out, open upper windows on side away from wind. Using windows in opposite walls gives best effect, but perpendicular walls can serve well. For normal velocity, windward openings (intakes) should equal 10 percent of floor space; outlet openings should equal 12 to 13 percent.

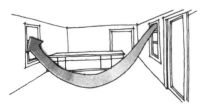

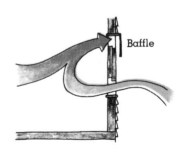

Baffle

Low velocity ventilation. Open windward windows much wider than downwind ones.

High velocity ventilation. Open windward windows slightly, downwind side much wider.

Single wall ventilation. Sometimes, the only way to ventilate a room is to establish a "convection loop"—inlets and outlets for air in the same wall. First shelter the upper opening so wind pressure does not stop escape of heated air.

About fans

Though natural ventilation is ideal when conditions are right, a cooling flow of fresh air through the house can be guaranteed by use of fans. Often fans provide cool indoor temperatures for less energy than air conditioners would use.

Fans can be grouped into three kinds: circulating, ventilating, and exhaust. All are rated on the basis of their ability to move air as measured in cubic feet per minute (cfm). But the system is not even close to exact, partly because not all manufacturers measure the same way, partly because fans work differently.

Wattage is another measure to look at when buying a fan.

Circulating fans do what their name implies—keep air moving and mixing. They don't ventilate. The group includes those little squirrel-cage desk fans, oscillating fans, ceiling fans of the sort associated with old films about the tropics, and box or floor fans. The range is from smallest to largest.

Though circulating fans do not lower air temperatures much, if at all, their light to brisk breezes do produce a cooling effect by accelerating evaporation of moisture from the skin. Their air moving capacities are measured in varied ways, because some move air at high velocity in a channeled direction, while others waft broad swaths of it at a leisurely rate.

None uses much wattage, the range being 9 to about 500.

Ventilating fans set up currents of new air in the house by blowing out old air. They include window fans, attic fans, and whole-house fans.

Window fans are the simplest and least expensive type in this group. Their housings are designed to fit temporarily into standard window openings, and most plug into regular outlets.

Window fans replace absent breezes by expelling air, on what usually is the downwind side of the house with enough velocity to pull fresh air in through opened windows on the windward side. They have 13 to 20-inch blades and draw 150 to 400 watts.

Large attic fans—permanently installed in a gable—do

Circulating fans

Oscillating fans, and box or floor fans share these virtues:

• Completely portable
• Need not be installed, only plugged in
• Modest purchase price and low operating cost

Oscillating fans, usually with 9 to 12-inch-diameter blades, move a considerable amount of air. Box fans, with 16 to 20-inch blades, move so much air you have to nail down your paperwork 10 feet in front of them. The latter move 8,000 to 12,000 cubic feet of air per minute, more than twice as much as typical oscillators.

These figures point up the wide variations in ratings of fan capacity to move air, especially when compared to the maximum 2,300 cubic feet per minute noted for the far more powerful whole-house fans. Part of the discrepancy lies in the fact that whole-house fans are measured against the air volume of the entire building, while circulating fans are more often measured against the air volume of an average-size room.

In any case, the job of oscillating and box fans is to move air a short distance at high velocity to produce a cooling sensation, in addition to stirring all the air in a room.

On the latter point, the closer a fan is to the floor, the better its effect because it then has the best chance to stir the room's coolest air.

Again, the measuring stick of cubic feet per minute poses as many questions as it answers. However much air is moved, the velocity is low, so the immediate cooling effect is rather less than with oscillating or box fans, but so is the nuisance factor of gusty winds.

Many contemporary designs have variable speed switches, permitting fine control.

Primary energy-saving features to look for are a favorable watt to cfm (cubic feet per minute) rating, a three-speed switch, and nonsymmetrical blades.

Ceiling fans are a different case; they must be installed. However, this can be as simple as exchanging a ceiling fixture. Purchase prices do tend to run higher than for other circulating fans. Sheer romance makes up for these deficits, though, and the gentle breezes are beyond compare. Blade diameters range from 30 to 56 inches. The big ones are said to move 5,000 cfm of air at a lazy 100 rpm and twice that much at 220 rpm. Ceiling fans draw as few as 9 watts and seldom more than 80.

Oscillating fans

Window fans

Desk fan

Ceiling fan

much the same job, but add a chimney effect as well. They must be wired. In most houses a ceiling opening must be cut to allow house air to flow upward into the attic.

Whole-house attic fans take the idea of ventilation to its limit. Installed in the ceiling or attic gable, they are designed to pull large amounts of air through a house and expel it from the attic.

The most powerful of the ventilating types (their ratings run from 1,000 to 2,300 cubic feet per minute), whole-house fans must be chosen and engineered into a house with care. Though other fans tend to be within the skills of weekend carpenters, whole-house fans frequently demand a professional installer and an electrician.

Exhaust fans, unlike ventilating fans, are not meant to set up currents of new air. Their role is mainly to vent odors, heat, and moisture to the outside from kitchens and baths. Most exhaust fans are designed to move small amounts of air from a confined space. If not able to do the complete job, at least they make easier work for larger fans by leveling the extremes of temperature.

One of the most common of the exhaust types, the kitchen fan typically makes 15 air changes per hour while running. Measured in about the same way as whole-house fans, its rated capacity is 60 to 200 cubic feet per minute. Wattages seldom exceed 150.

Most kitchen fans are installed 21 to 36 inches directly above the range. The shorter the ducting, the more efficient the air removal; this means that installing directly through a wall is optimal. There are both squirrel cage and regular bladed kitchen fans; both types are about equally effective.

Bathroom fans, similar in purpose and design to kitchen fans, also are sized to make 12 to 15 complete air changes per hour. Usually they are mounted high in an exterior wall, or in the ceiling. (If ceiling mounted they must be ducted outdoors to avoid excess moisture in the attic.)

Many codes call for the bathroom fan to be wired so it turns on automatically with the light, a nuisance for those who wish to retain indoor moisture during the cold of winter.

Window fans

Fans that can be installed in an existing window provide the most effective ventilation available at their low cost.

The installation of a window fan is quite simple in the standard sash windows they are designed to fit. The only work necessary is to provide a good base for the fan to rest on. It is also important to seal the portion of the window not occupied by the fan.

Many window fans are manufactured with adjustable slides that open to fit the window width precisely. If you have a model that doesn't have these slides, you may wish to make a weathertight installation.

Cut a piece of ¼-inch plywood to fit the window opening, then cut an opening in the plywood for the fan. A foam or felt gasket will insulate and cut down vibration noise. You can secure the plywood in the window frame by tacking strips of molding against it at each side. (Remember that you may wish to remove the fan during the cold seasons.) Finally, the plywood can be primed and painted for appearance and weather protection.

The fan should be installed in a window on what is usually the lee side of the house (the direction away from most wind). The fan, of course, should be positioned to exhaust air to the outside.

Because the unit lets air pass through freely, it should have its grille covered during extremely hot or chilly weather.

Grille covers can be bought, or made from pliable materials.

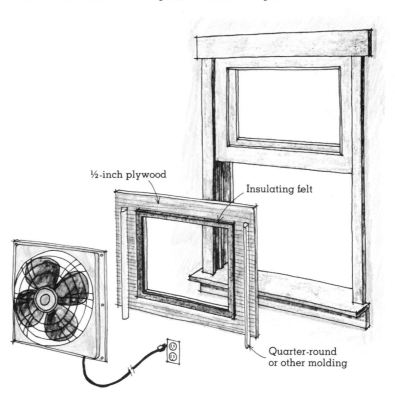

½-inch plywood

Insulating felt

Quarter-round or other molding

Whole-house fans

What the various models of whole-house fans have in common is a fan assembly with blades 24 to 36 inches in diameter, a motor with drive belt to turn the blades, and a chassis to hold the assembly together.

The basic unit does not include shutters that close the intake and exhaust vents when the fan is off, nor does it include switches and wiring.

Installation requires considerable skill in carpentry and wiring.

Whole-house fans can be installed horizontally in a ceiling, vertically above the ceiling (if a housing is added to direct air to the unit), or vertically in an attic gable. No matter which of the three, the fan requires cutting through the ceiling and an attic wall for vents. Both openings need to be framed; any cut severing one of the house's structural supports (a joist, stud, or rafter) requires headers (cross supports) to be fitted across the severed member. Single headers will be stout enough to bear the modest weight of any fan of this type, as shown, but you can be cautious and install doubled headers.

Wiring can be in one of several designs. The switch can be mounted on the chassis and operated with a pull chain, or wired as a conventional wall switch. If wired as a wall switch, single speed units can be linked to a thermostat or timer. If the fan is a variable speed model, the switch can be a rheostat, which allows variable amounts of current to pass.

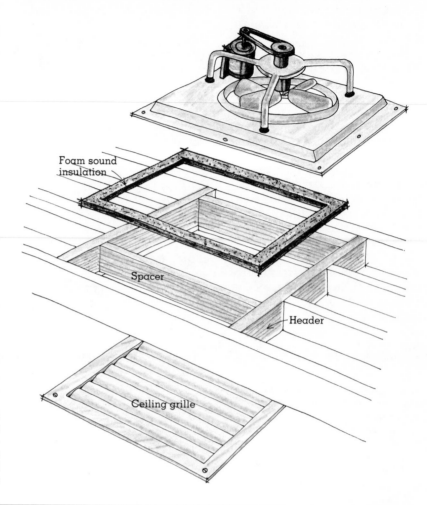

Foam sound insulation

Spacer

Header

Ceiling grille

Installing exhaust fans

Kitchen and bathroom exhaust fans are installed and wired in much the same manner as whole-house fans, but because the openings are smaller, structural members seldom require cutting.

If the fan is set into a wall, a single opening is all that is required. Most units are small enough to fit between two studs, though obstacles like built-in ranges and immovable gas pipes sometimes force the fan to be located on the face of a wall stud or ceiling joist. If the cut falls between structural members, you may or may not need to toe-nail nailing blocks in place to secure the vent; manufacturer's instructions will specify. If you must trim a stud or joist to accommodate the duct, you

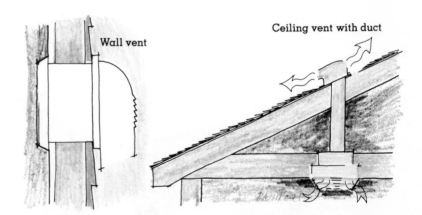

Wall vent

Ceiling vent with duct

will have to nail headers in place as described above for adding a whole-house fan.

If the fan is set into a ceiling, airtight ducting must extend from the fan housing to an outside vent in order to avoid excess moisture or the risk of grease fires in the attic.

Either type of fan may be operated with a pull chain; most are wired to a switch in the wall or in a range hood.

Attic ventilation

Proper attic ventilation is a key to developing a truly comfortable, energy-efficient home. In a poorly vented attic, heat trapped on a hot summer day can reach 150°. The result is either an uncomfortably warm house or the need to support an air conditioner that labors long hours and uses excessive amounts of costly power.

A well-vented attic will keep excessive heat from building up in the summer months and help control humidity in winter. The costs of attic ventilation are modest and the solutions virtually permanent.

There are two basic methods of attic ventilation. Natural ventilation uses a system of vents that depend on natural wind pressure and thermal air movement. Power ventilation uses ventilating fans (discussed on page 14) to push hot or moist air out through roof or gable vents.

Attic fans are installed in an attic wall near the peak of a gable or directly through the roof as close to the peak as is structurally feasible. They can function efficiently using thermostat or humidistat controls. Though the hardware for attic fans costs more than for natural venting, the fans usually require less structural work (fewer openings to be cut) and can be less costly to install, particularly if you're having the work done by an outside contractor.

To determine the size of fan or number of vents you need, follow the general formulas based on the number of square feet in your attic floor:

A ventilating fan should have the capacity to move from 1½ to 2 cubic feet of air per minute (cfm) for each square foot of floor area. For example, if the floor area of your attic is 1,000 square feet, the fan should be capable of moving from 1,500 to 2,000 cfm. At a ventilating rate of 1½ cfm (per square foot of floor space), on a hot day, attic temperatures can be reduced about 45 percent. At 2 cfm, temperature reduction could be as much as 67 percent. Tests of power ventilating systems have shown that an air flow rate higher than 2 cfm (per square feet of floor) will not produce significant additional temperature reductions.

In a natural venting system, 1 square foot of clear vent space is required for every 150 to 250 square feet of attic floor area. Consider in your calculations any screens, louvers, or grillwork placed over vent openings, as they slightly reduce the flow of air (your building supplier may be able to tell you by how much).

The placement of vents in a natural ventilation system is critical.

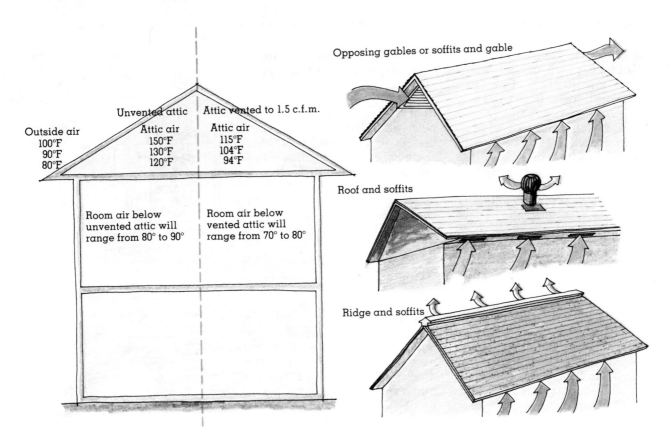

Outside air
100°F
90°F
80°F

Unvented attic
Attic air
150°F
130°F
120°F

Attic vented to 1.5 c.f.m.
Attic air
115°F
104°F
94°F

Room air below unvented attic will range from 80° to 90°

Room air below vented attic will range from 70° to 80°

Opposing gables or soffits and gable

Roof and soffits

Ridge and soffits

At top left and top right are variations in gable vents; lower left is a wind turbine, lower right a roof vent.

Gable vents. These work most effectively when the wind blows toward them. One in a gable wall away from wind can be coupled with soffit (eave) vents, which bring cool air in at a lower point. More usually there are two gable vents, one toward wind and one away.

All gable vents should be installed as near the roof ridge as practicable, because the hottest air gathers at the peak.

Where natural venting action cannot do the job, attic fans are the only solution (see discussion on page 14). Some fire codes prohibit them, though.

Gable vents do not require much carpentry. Since gable walls do not bear structural weight, cuts and framing are simple and straightforward as shown. Roof vents demand more skill to avoid leaks later.

Remember to cover the inner side of static vents with screen-ing to keep out insects, rodents, and other unwanted small visitors.

Roof vents. Wind turbines are most effective where summer heat is accompanied by at least some breeze. The turbine's slotted, free-turning ball creates an ever stronger vacuum with increasing speed, and sucks out correspondingly greater volumes of air from the house. It is equally effective no matter which direction the wind blows. When it loses its vacuum effect in still air, it ventilates only by natural thermal action.

Lower profile roof vents depend heavily on the fact that wind action creates a slight negative pressure on the downwind pitch of a roof. This boosts the simple rising of heated air. As a result such vents work best when placed on the downwind pitch.

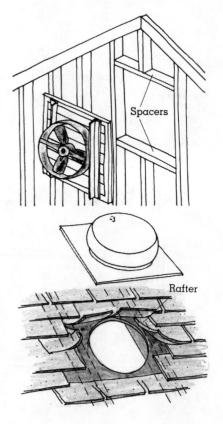

Spacers

Rafter

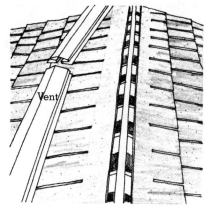

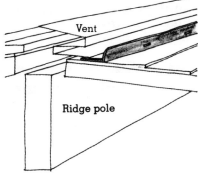

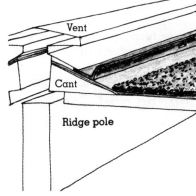

Ridge and soffit vents. When coupled with soffit (eave) vents, continuous ridge vents are the most effective of natural types for three reasons. First, they are directly at the ridge, so hot air can continue to rise with or without a helping breeze. Second, their length produces even ventilation throughout the attic. Third, the presence of a vent channel on each side of the unit means that it always has both a windward and a downwind side, whatever the wind direction, and thus can always vent toward negative pressure.

In themselves, continuous ridge vents are relatively easy to install. As shown, ridge cap shingles are stripped away; then the roof sheathing is cut away to a width of 1½ inches on each side of the ridgepole.

On flat or nearly flat roofs, a triangular wedge strip or cant (see drawing) must be nailed on each side of the opening. The lengths of ridge vent are then nailed in place. Only one metal ridge vent section needs to be cut to fit the last gap.

The main obstacle is that these units must have regularly spaced soffit vents all around the eave line as air intakes. Depending on the material used to enclose the eaves, installing these can be difficult.

MAKING GLASS BEHAVE

In practical terms, every square foot of plain window glass in your home is useful when you want to gain sun heat, but a disaster for the energy bill most of the rest of the time.

Plain glass allows about 80 percent of the sun's radiant heat to pass into the indoors for as long as the sun shines on a bare window. When the outdoor temperature differs sharply from the one you want indoors, the 0.89 R-factor of glass means you gain or lose about 12 times as much heat per square foot of bare glass as you do through a completely uninsulated wall. (See "About R," page 49.)

Windows and glass doors are—obviously—crucial points to attack in rehabilitating a house that costs too much to keep comfortable. As with controlling air movement, the goal is to establish flexible control, so furnace heat can be kept in, and sun heat allowed in or kept out as the season demands.

The remedies for problem windows and glass doors cost more than caulking and weatherstripping, and they are more difficult to plan. The most effective ones compensate by paying for themselves in as little as a single season. Many compensate still more by offering some of the most architecturally attractive opportunities a would-be energy saver will find.

The main remedies for year-round control of heat gain and loss through glass must offer both insulating capacity and an ability to shade: interior or exterior shutters, shades, and draperies. Double glazing and tinted glass can contribute.

If you need mainly to avoid summer heat gain, the possibilities expand to include outdoor shading devices of every sort, from patio roofs and freestanding screens, to vines and trees. Tinted glass and window film also are possible shading solutions.

Though one study suggests that hanging a vinyl roller shade in every window can cut fuel bills by 60 percent if you use it properly, the hard question is to define "properly."

Orientation of windows to the sun, seasonal temperature extremes, and even winds all have strong effects on how and when you shade a window, and when and how much you insulate it.

The roles of the window as ventilator, viewpoint, and source of light have much to do with your choice of materials and how you use them. You may even find that you want to add to the amount of heat you gain through glass during the cold months by including a solar greenhouse or other passive solar device in your plans.

Heating and Cooling Savings through Covering Windows

Percentages in the chart assume the house is weathertight and moderately well insulated.

Action	Range of savings	Your potential savings
Daytime shading to reduce solar heat gain by 60% (See pages 22–23, 29)	In regions of high daytime temperature and cool overnight temperature, exterior shading of sun wall windows (especially east and west-facing) can reduce air conditioner running time by about 45%. In regions where overnight temperatures remain high, savings are much less because of heat gain at other points.	
to reduce solar heat gain by 80% (See pages 22–23, 29)	With same limitations as above, maximum savings on air conditioner use are about 63% with exterior shading, about 53% with interior shading.	
Nighttime covering with simple vinyl shade (See page 22)	Can reduce heater's annual energy use by about 1% according to consensus.	
with thermal shutters or draperies (See pages 22–27)	Savings on energy use by heater can be about 7% a year at R-5, 12% at R-11. Savings on energy use by air conditioners are about 4 to 5%.	

Following the sun

Chances are your home is not oriented true to the compass. Still, these general descriptions can help you think about how much heat you gain or lose from each side of your house, and where to start correcting any problems.

North windows. These never gain sun heat but can lose great amounts when outdoor temperatures are lower than desired indoor ones. They are good candidates for thermal coverings in severe-winter areas.

East windows. Frequently useful for heat gain in all seasons, these rarely gain too much heat. They may require some covering against heat loss after sundown in severe-winter regions.

West windows. These are a problem. They gain too much heat in summer, and, in winter, they gain too little to compensate for what they lose. They require shading in all climates and may require thermal covering.

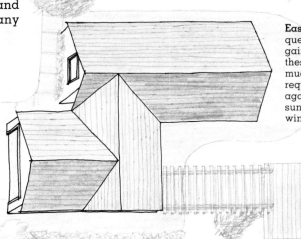

South windows. A homeowner's best friend, these provide long daily exposure to sun at generally favorable angles; heat gain is controllable by simple shading. These are best windows to develop for sun heat. If extensive in area, they'll require thermal covering for severe winter nights.

The Sun's Seasonal Tracks

These charts show how many BTUs (British Thermal Units) you gain per hour, per square foot of unshaded glass in each of the four seasons. Remember in measuring the gain that an average-size gas furnace running constantly delivers only 80,000 BTUs per hour.

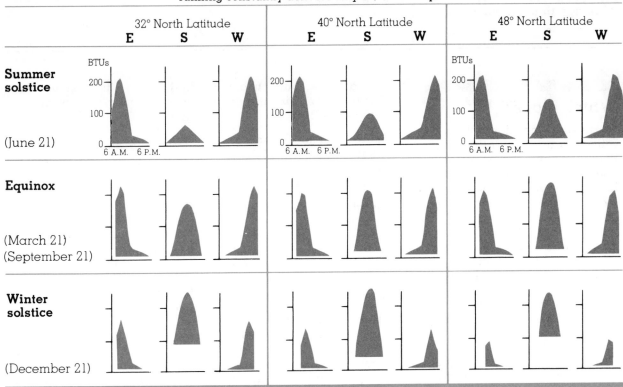

Window coverings

The virtue of inside window coverings is their flexibility rather than their versatility: they can be folded or rolled, but few can function as both shade and insulation. Flexible indoor window coverings can be among the least expensive solutions to problem windows. They are designed for subtle adjustment as conditions vary, and they are easy to control. This last is perhaps their greatest asset and is why they tend to be used more regularly and effectively than other materials with greater inherent capacities but more troublesome patterns of use.

They are not, as the following descriptions show, interchangeable, or equal in effectiveness. Some are best suited to shading; others perform better at keeping furnace heat inside. Perhaps only multiple-layer see-through films can do both jobs well. This suggests pairing two materials to gain year-round control over heat exchange through windows.

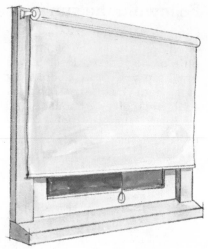

Venetian blinds. These are among the least efficient energy savers, because they transmit both sun heat and convection-warmed air. Still, because they maintain outdoor views and adjust light levels so well, they remain attractive and useful.

On an average, they reflect about 40 percent of the sun's rays; potential R-value is less than 0.5. At this efficiency, in a maximum-use test, they trimmed fuel use about half as much as a vinyl roller blind. Special designs with one reflecting and one heat-absorbing side are more efficient.

Vinyl blinds. Surprisingly effective sun shades, they reflect about 60 percent of the sun's heat. Potential R-value is about 1, though the surrounding frame must help. There must be a box valance at the top and channels or guides at the sides to form reasonably tight seals between shade and frame, or insulation value is less than the potential.

Pluses: They are inexpensive to buy and easy to install. Drawbacks: Closed, they eliminate views outside and darken the room greatly.

Which glass?

Glass varies considerably in its reflecting and insulating qualities. Knowing the qualities of your window glass may affect how you cover it, or may even lead you to replace it with another type.

Float glass. This is common window glass; it comes in $\frac{3}{32}$-inch or $\frac{1}{8}$-inch thicknesses. Either way it has an R-value of 0.89, and transmits about 80 percent of all radiant heat from the sun.

Plate glass. As far as energy efficiency is concerned, it differs from float glass only in being thicker (usually ¼-inch), which does increase R-value.

Tempered glass. Though it is stronger than float or plate, and safer when broken, its thermal resistance is about the same.

Safety glass. A layer of plastic is sandwiched between two layers of glass—the familiar formulation of car windshields. It has a slightly higher R-value than plate glass of similar thickness.

Heat-absorbing glass. Gray, green, or bronze, this glass varies in degree of opaqueness. The coloring materials absorb solar heat, then reradiate it to both sides. Its best use is in cool-summer/cold-winter regions.

Reflective glass. A mirrorlike coating of transparent metallic oxide is fired on the surface of ¼-inch plate to reflect up to 70 percent of solar heat and cut light transmission by as much as 50 percent. Its best use is in hot-summer/mild-winter regions. Conventional glass can be made reflective by application of commercially available polyester films.

Patterned glass. Passing glass through patterned rollers creates ribbing, pebbling, and other surface effects that cut

Films. These come as metalized polyester or mylar on a spring roller, or as semirigid polyester or plastic secured to the window frame as a sort of storm window. The ultimate roller type has five layers of mylar designed to separate so air pockets between each layer open to produce a potential R-value of 13 when properly installed.

Colors and reflectance values (see Glossary, page 94) range widely; most films are smoky or bronze, with reflectances of 60 to 80 percent. An advantage over vinyl is see-through quality.

Insulating shades. As fuel costs rise, ever-thicker roll-up shades are being produced to insulate windows.

Though they can be used as 100 percent reflective sun screens, they blacken rooms. They therefore serve best in lieu of thermal shutters during dark hours (see "Shutters," page 24).

The thickest ones can have an R-4.25, if they have a box valance at the top and guides at the sides to seal air between them and the window. Some of these shades come with guides included.

Draperies. Draperies that are potentially very efficient have insulating foam lining or foil. Both insulations come either as integral parts of the draperies or as zip-in, zip-out liners.

Depending on fabric, unlined draperies reflect 50 to 90 percent of sun heat. Potential insulation value ranges from insignificant to R-1 in unlined draperies, and to R-3 in insulation-lined ones. The latter must fit tightly against the window and floor to achieve their insulation potential. This usually means a valance at the top, side guides, and a weighted hem below.

---

down transmission of solar heat to some degree, though much less than reflective glass.

Insulating glass. Also known as thermal glass or double glass, it actually is two pieces of float or plate glass separated by an insulating dead air space. If clear, it allows the same radiant heat gain as a single-glazed window, but has an R-value of 1.81 (when the air space is 1 inch or more).

Triple glazing increases the thermal resistance to R-2.79, but its high cost makes payback improbable.

See "Storm windows," page 29, for a quick, inexpensive alternative to double glazing. The potential difficulty in older homes is windows with fancy sills or other ornaments; the storm windows must fit against a flat surface to be effective.

Sashes and casements

Sashes and casements are the wood or metal units that hold window glass in place; these units in turn fit into the frames of openable windows.

In addition to variable airtightness, they resist heat transfer sometimes well, sometimes badly.

Most wood sashes or casements are about an inch thick, so have an R-value close to 1. If not designed to resist heat transfer, metal units have a much lower R-value, but those designed with a gap—called a thermal break—can be more efficient than wood. The absence of a thermal break is easy to see; the two pieces of metal gripping the glass fit tightly against each other, while sashes with thermal breaks have a visible separation between the two pieces of metal.

If you're troubled by sashes that conduct too much heat, an expedient solution is the clear nylon or vinyl sheet noted under "Reflective glass" (preceding). Take a close look at your window frame before buying one of these, to be sure you find one that will fit tightly.

Shutters

Shutters are enormously efficient window coverings; they're also blasted nuisances.

Their flexibility of design allows shutters to give infinitely fine control over heat gain from sun radiation—from none at all to as much as the window allows.

Shutters can shade and/or insulate. Insulating types are called thermal shutters. Where extreme heat or cold is a problem, specific thermal designs can change a window's R-factor from 0.8 to 11.0 (see "About R," page 49).

Shutters do not need to cut off ventilation when it is desirable. They do not even need to diminish light to unusably low levels. Finally, they are diverse enough in appearance to fit almost any style of architecture.

But to do all this they need to be adjusted daily, perhaps several times a day. This tends to keep exterior shutters from being as popular as indoor ones, in spite of the outdoor type's slightly greater efficiency as a sunshade.

The sketches on these pages show most shutter variations; the accompanying text points out some of their pros and cons. Here, briefly, are the questions to ask:

• How much will reradiation of heat from the shutter into the room be a factor in hot-summer climates?

• Is there a need for the shutter to provide insulation in winter?

• What effect will the shutter have on summer ventilation?

• How will it affect light and view?

• Will the shutter be easy to operate if outside a window of any given operating design?

• How easily can the shutter be secured to a window or wall?

• Should it offer security if windows are left open at night?

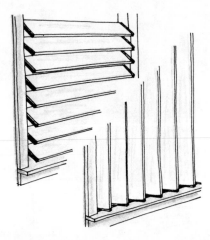

Adjustable louvers. Best installed outside, large-scale louvers can be set vertically or horizontally. The choice often is governed by available views (vertical louvers often let you see less unless you sacrifice shade). The type often can be bought as a pre-fabricated metal unit; it also can be made using wood for louvers and metal tracks much like those in jalousie windows.

Bahamas. For outside use only, and only for sun control. Partially raised, they can provide full shade while permitting excellent cross-ventilation, so are popular in hot, humid climates. Most have fixed small-scale louvers that permit some light to enter, and some view to the outside. A few have adjustable louvers. The type is available in aluminum and wood. It offers some security.

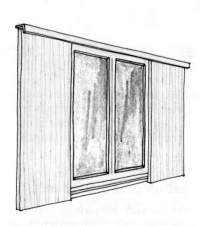

Sliding shutters. At their most useful inside when furnishings crowd against a window, or as covers for sliding glass doors. With doors, they are easier to maintain if top-hung; floor tracks tend to collect clogging dirt. Can be of varied materials to suit shading or insulating needs. For maximum insulation value, must fit tight against window frame, so should meet, not be offset.

Side-hinged shutters. This is the classic image, especially when of wood with small-scale louvers. Though primarily for shade, even this type has some insulating value. Solid materials can be sandwiched together to form highly insulative window covers—though at expense of lost light and view. Can be installed inside or out, to almost any vertical size. Big ones must have heavy-duty hinges.

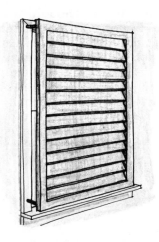

Combination bahama and side-hinged. An outer wood frame hinged at the top allows the whole unit to swing up for ventilation at no loss of shade. Standard side-hinged shutters fitted into this frame open to admit sunlight or allow view. Practical only for outside installations. While most have shutters with small-scale louvers, solid materials can be used for extra insulation value.

Fixed frame. For outside use only, and only for sun control. Most admit light through small-scale fixed louvers as well as at sides, permitting only a limited view but effective ventilation. One virtue of the design is good security; the frame fits close to the wall, so the window behind can be left open to cooling night air. Available in aluminum and wood to fit most standard windows.

Rolling shutters. These are installed only outside, but most are designed to be controlled from inside. Some have electric drives; some have hand cranks. The shutters are either adjustable louvers or interlocking screen or slats. They are available in aluminum, steel, PVC, and wood. Some slatted PVC units insulate to R-5, but most are purely shades. Security is a bonus factor.

Multiple fold shutters. A variation on the simple side-hinged type, bi-folds and multiple-folds lend themselves to spanning wide windows. Folds mean shutters need less space to swing, and allow flexible control of light. Bi-folds can support themselves on side hinges alone; multiple folds require a top track, too. These units can be installed outside, but most are used indoors.

Top or bottom-hinged. Outside, top-hinged types double as awnings; inside they are useful with long rows of closely spaced windows (as on page 26). Bottom-hinged types attached outside are often lined with foil so they double as reflectors of sun heat in cold winter regions. Main purpose is insulative. Hinged top or bottom, they tend to warp when opened often for long periods.

Pop-ins. Used only inside, and only to insulate, most are made of foam or batt insulation sandwiched between plywood or other covers. Beveled edges help achieve required close fit in the window frame; two opposed sets of cleats can help keep them in place where frame is not deep enough to do so. A bulky nuisance to store, but unbeatable insulators in severe winter climates.

Making your own

Most of the shading shutter types are readily available as manufactured units; the homeowner needs only to affix hinges or tracks, and latches, and the job is done.

Insulating (or thermal) shutters for severe climates are harder to come by as manufac-

tured items. Though several companies now make effective insulating shutters, this still is a place to exercise ingenuity.

Most window frames are not deep enough to accommodate shutters using batt insulation. The usual materials are insulating board with reflective foil sandwiched between the board and an inner covering.

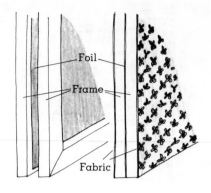

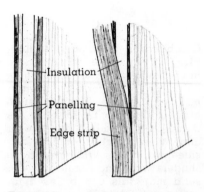

Typical homemade thermal shutters use core of insulation board, a sheet of foil, and non-flammable coverings. A wood frame improves rigidity, and is solid anchor for hinges. Shading shutters can be wood frames with a fabric sandwiched between laminated frame, or can have fabric attached to faces of frame. Foil liner helps insulation, but decreases light.

Sliding shutters belong not only over sliding doors, but over windows extending to floor (or almost to it) where hinged types would be restricted by nearby furniture.

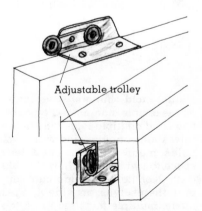

Adjustable trolleys and track for sliding shutters are easily installed in most window frames.

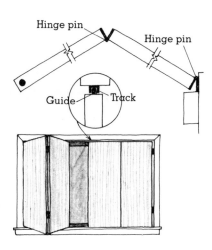

Folding shutters fit most window shapes, and can be made of a wide range of materials to suit diverse purposes; many designs can be bought as manufactured units.

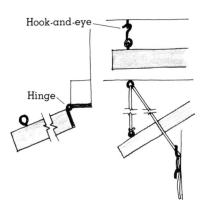

Hook-and-eye is good system for shutters within reach; higher ones need simple eye overhead and cleat on wall for cord.

Top-hinged shutters work well where several windows are in a row. They adapt well to installation on clerestory windows, or others out of reach overhead, can be hazards in traffic areas.

Hanging hinged shutters

Large shutters, because of their weight, usually are mounted with surface-mounted strap hinges.

The surest approach is to secure the straps to the window frame or wall, and then, with the shutter propped in exact position, drill guide holes and drive the screws on the shutter side of the hinge.

Lighter shutters usually are mounted with leaf hinges recessed into mortises. More often than not, leaf hinges have to be adjusted in their mortises at least once to secure a square, proper fit. For ways to make these adjustments, see page 11.

Remember, the hinge pin faces in the direction the shutter swings to open.

Step by step, the procedure:

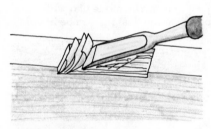

1. Prop the shutter in precise position. Mark hinge locations with knife cuts across adjoining edges of shutter and window frame.

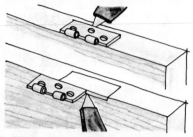

2. With a hinge leaf as a guide, use a knife to score an outline for each mortise, and also to mark the depth of the cuts.

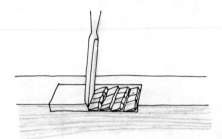

3. Make a series of parallel chisel cuts to desired depth. Hold chisel vertically with bevelled face away from you; tap with hammer.

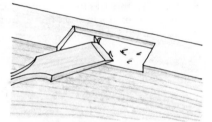

4. Decrease angle of chisel to about 30°, and chip carefully at loosened wood. Do not use a hammer to avoid splitting too deep.

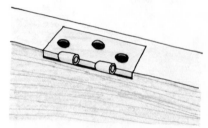

5. Make final smoothing cuts with chisel held almost flat; work from the side. At intervals place leaf hinge in mortise to check fit.

6. Take special care at corners so hinge leaf can rest flat. If a spot is too deep, cut a shim from hard-finished cardboard.

Installing adjustable louvers. Adjustable wooden lovers can be installed either in manufactured tracks much like those for jalousie windows, or in simpler and less costly homemade frames. In either case, the trick is to get the opposite tracks exactly level with each other and parallel in the window frame.

The manufactured tracks are designed so that all louvers move together when pressure is exerted against any one.

Vertically mounted units are slightly easier to install since you can install the top track, and then use a plumb bob to guarantee precise placement of the lower track in both planes.

To install a horizontal unit, secure one track using a carpenter's level, then use one louver and the level to locate the opposite track.

Attach the top end first, then repeat the check at the bottom before securing it. Only after both tracks are securely in place should all of the louvers be installed (see lower sketch for how to use level).

You may have to shim a track or trim individual louvers to compensate for a window frame not precisely square.

If you elect to make your own frame, measure the window with care and cut two side (or top and bottom) strips to which you will fasten the louvers. Lay out the louvers in closed position to be sure of correct overlap and enough edge clearance to allow the louvers to close tight. Assemble the unit on a flat surface, using pivot pins bought from a building supply store; then mount it complete (see page 34).

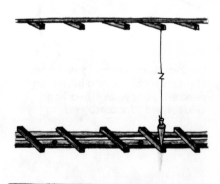

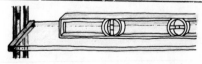

Storm windows and screens

In addition to all the types already noted, there remains a small galaxy of miscellaneous window coverings not meant to be adjusted every day, but rather put up for at least a season. Storm windows and screens are the notable members of the group.

Storm windows. As their name suggests, they are meant for winter use. Their twofold purpose is to cut down on air exchange, and at least double insulation at the window. Manufactured units using glass in metal frames work well but are expensive in relation to the amount of help they give. This fact has given rise to all manner of plastic sheeting and film devices, some manufactured, many homemade.

Homemade storm windows using lath with vinyl film, or other plastic film, work as effectively as the metal-and-glass manufactured unit. The homemade unit costs mere pennies, but it seldom lasts more than 2 years and is not handsome while it does.

A compromise midway between the two is a sheet of rigid vinyl or nylon with a patented device to attach it to a window frame. Some attach with the same gripper as cuffs on blood pressure sensors; others use snaps or similar mechanical fasteners; still others have an adhesive coating.

Storm windows can be attached indoors or out, depending on the location of the window in its frame. The goal is maximum airtight space between the regular and the storm window.

Screens. Most useful as shades in summer, they double as protection against insects in windows opened for ventilation. A surprising range of reflectances is available in screens (see reflectance in Glossary, page 94). They are made usually of metal, fiberglass, or nylon mesh, but the material appears less important than two other factors: the fineness of the mesh, and, less visible but more important, its exact design.

The most effective screen at publication—an outgrowth of research for the space probe called Skylab—reduces heat and glare by 70 to 80 percent while blocking only 38 percent of visible light. Traditional wire screens come closer to 50 and 50.

Screens are far more effective mounted outside than inside a window.

For a tight job, staple film to lath, then tack lath to window frame. It is also possible just to stretch the film across the window and tack lath over it.

Metal or wood cleats can secure plastic film on a lath frame to a window. The advantage: they do not leave an accumulation of nail holes in wood.

Conventional storm windows of glass in metal frames are both durable and sightly, but quite expensive relative to their overall effectiveness.

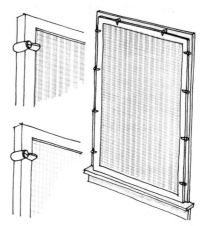

For summer shading, shop around among types of screening. Their performance levels as heat baffles vary; the type at top is efficient.

Vertical screens

Sun-struck windows—and even whole walls—can be shaded economically and with little difficulty using any of a number of vertical screen designs.

Where sun control is the only need, such screens often are more satisfactory than shutters: they allow reflected light to enter so do not demand daily attention. But this asset really applies only where views are not impressive enough to be in daily demand.

For most of the country, vertical screens should be flexible, removable, or both, to take advantage of sun warmth during the chilly seasons.

Flexible materials can be any commonly used for awnings or shades. Canvas, vinyl, bamboo, and roll-up lath all have been installed successfully. As the adjacent drawing indicates, plants are yet another possibility.

The ways in which vertical screens can be attached are at least as diverse as the materials. Sliding screens have been suspended from eaves on traverse rods, on rods with rings (like those used with shower curtains), and on wire strung through grommets fastened in the material. Roll-down screens have given rise to dozens of ingenious rope-and-pulley arrangements. Sliding and roll-down screens each have advantages, depending on your requirements and the relationship of window to garden. In general, roll-down screens permit more flexible control of light with a minimum loss of the view.

A key consideration in windy regions is the need to limit motion in the screen to keep it from destroying itself.

A few parts of the sun belt profit from permanent screens of either masonry or wood, attached or freestanding.

Roll-away planter. Where there's a will, there's a way. This roll-away planter provides a cool, green screen from the afternoon sun for a window that faces a large expanse of paved patio. When cooler weather arrives, it can be pushed away from the window or even hidden away completely.

Rough 1 by 10-inch redwood is ideal for the box. A frame of 1 by 2-inch redwood holds up wires for vines to climb. Heavy-duty 2-inch casters make the planter moveable. (Because of the weight of the soil, practical maximum length is about 5 feet.)

For the strongest assembly, nail the side panels to the edges of the bottom, and overlap the end pieces as shown. Use galvanized nails throughout. For long life, you may wish to use wood pressure-treated with a fungicide.

For drainage, drill six to eight 1-inch holes in the bottom.

For a list of usable plants, see page 38.

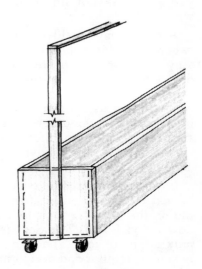

Matchstick blinds. Hung on traverse rods, these work more effectively outside than in, if they are protected from buffeting winds. Mounting the traverse rod well up in an eave protects it and the blinds from rain. A length of cable strung through eyes in 6-inch L-brackets can minimize wind buffeting.

The traverse rod and alternate means of securing blinds are shown below.

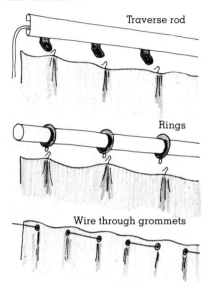

Traverse rod

Rings

Wire through grommets

Roll-down shades. These can be mounted under eaves or at the edge of a patio roof. The same rope-and-pulley system used for awnings or bamboo blinds can be adapted to canvas, vinyl, or other shade materials. The rolled-up blinds can hide behind the fascia at the edge of the eave, or between two overhead members.

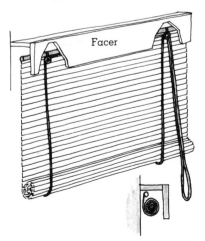

Facer

Masonry can make an effective permanent sun screen. Precast concrete screen blocks come in any number of designs and thicknesses. Such a screen must be erected within a frame much like that for a fence. See the *Sunset* book *Fences & Gates* for ideas and techniques.

Vertical louvers make excellent, highly adjustable shade screens. Like the masonry screens above, they are built much in the manner of fences. For other details on construction, see pages 28 and 34 of this book.

Overhead screens

Overhead screens are more troublesome to design and build than vertical ones, but they have incomparable advantages: they preserve views and expand outdoor living space. They also lend themselves readily to shading large expanses of glass or even solid wall.

Size options range from a modest brow over a south-facing window to a gigantic patio roof. In either case, the most practical approach requires built-in flexibility, so the pale winter sun can help warm the interior and the summer sun can be balked at overheating it.

Brows, awnings, and other small overhead shades should be designed with sun angles foremost in mind. The effect of even a modest extension of eaves or other overhangs can be surprising. The diagrams on page 21 show how much sun heat you can keep at bay by adding two or three degrees to the angle of shadow on a west or south window.

A substantial, permanent framework can be built to support removable shading. The shades can be of wood in various forms (spaced lath panels, adjustable louvers), or canvas, vinyl, or other pliable fabric stretched over a frame. Not least effective as shades are deciduous vines trained on an arbor.

Where the overhead covers a large expanse of patio, the angle of sunlight becomes irrelevant. The key to cooling is to allow heated air to escape upward, even when the overhead is providing maximum shade. Heat trapped beneath a solid roof can more than offset any benefits of shade.

Major patio overheads are subject to building codes and should not be built without required permits.

Sawtooth vinyl shade. Framework of 2 by 6s supports heavy, brightly colored vinyl to provide summer shade over deck located on west wall of home. Framework was designed to permit rain water to drain off to outside of deck. Vinyl can be rolled up and stored during winter season.

Landscape architect: Glen Hunt.

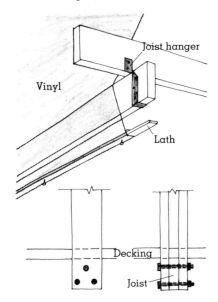

Wisteria arbor. Natural shading is provided by Japanese wisteria from early spring through summer. Arbor offers added bonus of an attractive flower display in spring.

The frame can be an extension of roof rafters as in the color drawing, or you can change the overhead's angle by plating from rafters, as shown below.

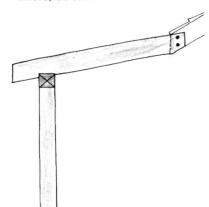

Louvers

Frame

Linked staples

Adjusting rod

Adjustable louvers. Adjustable louvers seem almost infinitely adaptable. They can be used as shutters and as fences, as well as overheads.

Correctly oriented, they can be fine-tuned from morning until late afternoon, not only to provide shade on warm days, but also to allow maximum penetration by the sun on cool but clear days.

For best sun control, the louvers should align north-south; that way they can tip east-west—toward or away from the sun as it prescribes its daily arc. If you have no opportunity to align your louvers true to the compass, make an effort to align them so they can keep as much late afternoon sun as possible off the area they cover.

This opportunity to fine-tune the amount of daily sunlight in an areas does not come easily. If the materials are weak or if the construction is not precise, the louvers may quickly become difficult or impossible to operate.

Most overhead louvers are made of 1 by 4 or similarly dimensioned wood. To avoid problems with uneven shrinking, swelling, or warping, it should be thoroughly seasoned before being cut to length.

Once the wood is cut to length, the next step is drilling holes in each end of each piece; into these holes will fit the pins on which the louvers will rotate. The holes must align with each other precisely for the louvers to rotate evenly. To save endless measuring, clamp together enough pieces to make a section, then use a straightedge to make a center line on the group.

After the louvers are drilled, drill corresponding holes in the two side rails that will hold them. Again, the holes must be located precisely. Space them to allow about ¼-inch overlap when the louvers are closed. The best bet is to drill both side rails while they are clamped together.

The holes must be drilled large enough to permit the louvers to turn freely, but those in the side rails should be small enough so the pins are held firmly. Pins usually are just galvanized nails.

Once the side rail pin holes are drilled, use a rail as a guide for locating brads in the wooden rod that connects all of the louvers in a section. Hammer the brads into place in the rod. Then lay the rod in place on an assembled section and hook a second brad through each brad in the rod. Hammer the second brads to the louvers as you go.

The sections then can be lifted into place and nailed to the overhead. This is a two-person job at least. Until nailed in place, the sections are rickety and difficult to maneuver; several small sections are easier to work with than one large one.

A simple awning. The absence of pulleys and cords makes this awning easy to install on a sun wall window. The upper rod rests in two U-brackets attached to the wall. The lower rod (of thinwall conduit) fits into two caps nailed to the wall. To avoid winds lifting rods out of caps, drill holes through conduit and caps, and insert nails as pins.

Lath panels. A far simpler (and less versatile) variation on the adjustable louver system shown on the facing page is modular lath panels that can be set into egg-crate overheads to create any pattern of shade desired. The panels need not be nailed in place; ledgers hold them securely, yet allow patterns to be changed easily to suit the changing seasons. Laths are spaced their own widths apart.

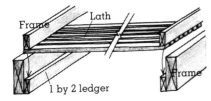

Frame

Lath

Frame

1 by 2 ledger

Thoughtful landscaping does not merely provide a cool garden; it cools the house in the process.

Cooling plants

Remember the last time you crossed a supermarket parking lot on a hot day and then went to a picnic on a shady lawn? The differences you felt between the two indicate how much landscaping can do to make a home more comfortable and to minimize energy expenditure. But achieving the maximum from your lot may require a great deal of personal energy.

A lawn is indeed more cooling for both you and your house than a concrete or brick patio, but shade is the real weapon against heat. Shade can come from either vines or trees, or perhaps from both. They provide an extra benefit by deflecting or redirecting winds. Block-ing winter winds can slow heat loss from the house. Channeling summer breezes toward the building can provide a welcome sense of cool.

Trees

The first and strongest impulse may be to choose trees we know and like, especially ones we loved in childhood. Generally these were very large shade trees.

But success depends on wisdom more than on emotion. The job is to understand exactly how you want the tree to control climate, and then to find the tree that will do the job.

The classic image of the farmhouse, snugged into a mixed grove of trees, is a picture that reaches much farther back in time than either air conditioners or insulation. A close look at the picture reveals that, in most cases, trees by the north side of the house are evergreen, those on the south and west, deciduous.

Evergreens planted close to a north wall form a pocket of still air, diminishing the ability of winter winds to steal heat. Deciduous trees to the south or west block summer sun. A dense tree overhanging a house can keep its temperature 20°F/11°C lower than it would be if unshaded on a hot summer's day. In winter, bare branches let a feebler sun warm the same building.

Basically trees offer three kinds of control: widespreading shade, a smaller overhead canopy, and vertical screening.

In the listings that follow, "D" stands for deciduous, and

"E" for evergreen. The numbers indicate climate zones defined on the map.

Wide-spreading shade. Usually this should be provided by a tree that will permit a lawn beneath it and that will grow enough to shade both house and garden, if desired. It should have a high foliage canopy, and should begin to give some shade in the second or third year after planting.

Maple (*Acer*) D 1–6 by species

Black locust (*Robinia pseudo-acacia*) D 1–6

Honey locust (*Gleditsia triacanthos inermis*) D 2–6

Tulip tree (*Liriodendron tulipfera*) D 2– 4, 6

Maidenhair tree (*Ginkgo biloba*) D 2–4, 6

Chinese elm (*Ulmus parvifolia*) D or E 6

Walnut (*Juglans*) D 2–6

Chinese pistache (*Pistacia chinensis*) D 4–6

Camphor tree (*Cinnamomum camphora*) E 6

Carob (*Ceratonia siliqua*) E 5, 6

Scholar tree (*Sophora japonica*) D 2–6

Linden (*Tilia*) D 2–4, 6 by species

Canopy-like shelters. Trees for this purpose should be well-behaved, giving minimum litter throughout the year. They should grow predictably to a limited size, if they are to give shade without dominating the garden. They should not require difficult pruning to give headroom.

Red horse-chestnut (*Aesculus carnea*) D 2–4, 6

Eastern redbud (*Cercis canadensis*) D 2–6

Flowering dogwood (*Cornus florida*) D 2–4, 6

Flowering cherries D 2–4, 6

Flowering crabapples D 1–6 by species

Flowering plums D 1–6 by species

Goldenrain tree (*Koelreuteria paniculata*) D 2–6

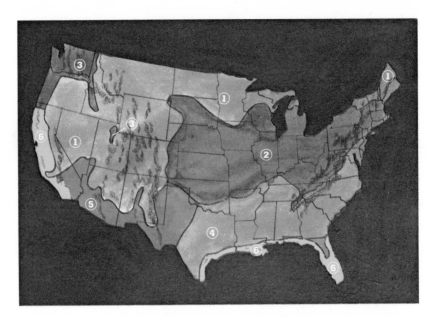

Plant climates. Zones on map indicate climates of varying winter severity. Matching listed plants with their recommended zones gives some indication of plant hardiness. Zones: 1. Mountains, range and basin country, upper Great Plains, northern Great Lakes, and New England; 2. Northeast, Middle Atlantic states, Ohio Valley, lower Great Lakes, Central Plains; 3. Northwest, Columbia Basin, Wasatch Front; 4. Southeast, Deep South; 5. Southwest deserts; 6. California and Florida.

Sourwood (*Oxydendrum arboreum*) D 2–4, 6

European mountain ash (*Sorbus aucuparia*) D 2–4, 6

Silk tree (*Albizzia julibrissin*) D 3–6

Modesto ash (*Fraxinus velutina glabra*) D 3–6

Glossy privet (*Ligustrum lucidum*) E 3–6

Chinese flame tree (*Koelreuteria formosana*) D 6

European olive (*Olea europea*) E 6

Persimmon (*Diospyros kaki*) D 2–6

Strawberry tree (*Arbutus unedo*) E 3–6

Vertical screening. For the purpose of directing wind or sheltering an area from wind, a tree with low or weeping branches is more effective than one that grows with a clean trunk. As wind screen, it should be 12 feet or higher. It works best if it filters rather than blocks the wind. Its width at bottom should not take away from gardening space.

Conifers (pines, firs, hemlocks) E 1–6 by species

Sweet gum (*Liquidambar styraciflua*) D 2–4, 6

European white birch (*Betula pendula*) D 1–4, 6

Pittosporum E 5, 6 by species

Eucalyptus, low-growing E 5, 6 by species

White alder (*Alnus rhombifolia*) D 2–4, 6

Poplar (*Populus*) D 1–6 by species

Vines and shrubs

Vines and shrubs are especially valuable interceptors of the hot afternoon sun: near the house or along a patio, they give cooling shade. Used as a second wall for the house, they can also act as an effective heat insulator. Trained on an overhead trellis, vines can substitute for canvas awnings or lath overheads. In all cases vines and shrubs are humidifiers.

Here are some things to remember when you use plants for weather control: A plant grown directly on the wall must have higher heat resistance than one trained away from the wall. A vine trained on a trellis built out from the wall of the house will be healthier because of greater air circulation.

As with trees, deciduous material works best where excess sun heat is only seasonal; evergreens are the answer where it is a year-round problem.

As you shop for plants to block the sun, you will become aware of the density of leaves. Your needs will determine the choice between dense shade, filmy or open shade, and a tracery of shade.

These vines and shrubs are old standbys as sun shades ("D" means deciduous, and "E" evergreen):

Wintercreeper (*Euonymus fortunei* and varieties) a vining shrub E 2–4, 6
Flowering quince (*Chaenomeles lagenaria*) shrub D 2–6
Kiwi (*Actinidia chinensis*) vine D 3, 4, 6

Wisteria vine D
Boston Ivy (*Parthenocissus tricuspidata*) vine D
Virginia creeper (*Parthenocissus quinquefolia*) vine D
Firethorn (*Pyracantha*) shrub E
Evergreen pear (*Pyrus kawakami*) shrub E
Star jasmine (*Trachelosperum jasminoides*) vine E
Evergreen magnolia (*Magnolia grandiflora*) tree E
Ivy (*Hedera helix*) vine E

These and more specialized vines are noted below with the drawings showing specific solutions to shading.

Window awning
For, cool, green shade in summer and interesting branch patterns open to sunlight in winter, kiwi vine from the basic list, and
Clematis jackmanii,
Chilean jasmine (*Mandevilla suaveolens*),
potato vine (*Solanum jasminoides*),
Solanum wendlandii,
Pandorea jasminoides.

Overhead canopy
For light structure and light-to-medium shade, kiwi vine from the basic list, and
Clematis jackmanii,
Guinea gold vine (*Hibbertia volubilis*),
potato vine (*Solanum jasminoides*),
Akebia quinata.

For sturdy structure and dense shade:
Clematis montana rubens,
Polygonum aubertii,
Wisteria sinensis,
Bougainvillea 'Barbara Karst,'
Clytostoma callistegioides,
Burmese honeysuckle (*Lonicera hildebrandiana*).

Terrace wall
To insulate against summer sun and to block wind, use a deciduous species. To block winter sun, choose an evergreen. From the basic list, choose wisteria. Also:
Clematis armandii,
Carolina jessamine (*Gelsemium sempervirens*),
Guinea gold vine (*Hibbertia volubilis*),
blood-red trumpet vine (*Distictis buccinatoria*),
violet trumpet vine (*Clytostoma callistegioides*),
Chilean jasmine (*Mandevilla suaveolens*),
Pandorea jasminoides.

Inner fence
To stop wind and trap heat, use dense evergreens adapted to pruning and clipping, including ivy and star jasmine from the basic list, and
Burmese honeysuckle (*Lonicera hildebrandiana*),
Clematis armandii,
plain or variegated Algerian ivy (*Hedera canariensis*),
Carolina jessamine (*Gelsemium sempervirens*).

West wall shade
For ground-to-eaves shading against sun in hot summer climates, with space between barrier and house for air circulation, choose wisteria from the basic list, and
hop (*Humulus japonicus*, annual; *H. lupulus*, perennial),
Carolina jessamine (*Gelsemium sempervirens*),
Burmese honeysuckle (*Lonicera hildebrandiana*),
blood-red trumpet vine (*Distictis buccinatoria*),
queen's wreath (*Antigonon leptopus*),
Campsis tagliabuana 'Mme. Galen.'

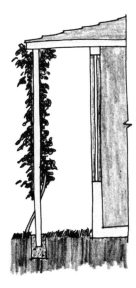

Permanent solar greenhouses can be attractive additions; what sets them apart from ordinary greenhouses is not appearance but details: a thermal mass, and special venting into the house.

The solar greenhouse

All greenhouses are in effect "solar." But today, with the increasing effort to utilize the sun's energy, greenhouses are being designed specifically to provide supplementary home heating.

A solar greenhouse differs from traditional greenhouses in that it incorporates extra mass for heat storage, is double glazed, and is built onto the south wall of a house.

Ideally, a solar greenhouse should be included in the design of a new home. Yet, if its heat can be vented efficiently to the indoors, a greenhouse is a practical addition to an existing home as a solar boost to a conventional heating system. These additions are feasible projects for homeowners with some building skills. The basic structure can be ordered from one of several greenhouse manufacturers or made at home. Many of the most workable ones now in service came largely from salvaged materials.

There are two basic requirements that must be met for a solar greenhouse to work. First, the house must have a wall facing within 30° of the true south and structurally capable of forming the north wall of the greenhouse. Second, there should be nothing blocking the sun when it dips to its lowest winter path. Deciduous trees are not necessarily obstructions, for they lose leaves in the winter and permit the sun to shine through. They can provide desirable shade during hot summer months.

How it works. A common demonstration of the "greenhouse effect" is provided by a closed car parked in the sun, winter or summer. Energy in the form of sunlight pours through the glass windows, is absorbed by the interior of the car, and remains trapped inside. Little warm air can escape, so temperatures inside can climb to 130°F/55°C.

This basic principle is what makes a solar greenhouse function. It absorbs heat, but the hot air produced is vented directly into the attached house to supplement its regular heating system. This, of course, reduces energy use and the monthly utility bill.

The cost. As is true with most solar heating systems, the more heat the greenhouse is expected to produce, the greater the initial investment. For example, if the most rudimentary solar greenhouse, built of common, inexpensive materials, can produce 20 percent of a given home's heating needs (the percent can vary greatly with climate and other factors), a solar greenhouse that provides 25 percent could take double the investment.

This general rule holds true for most efforts to reduce energy consumption. The first steps at weatherstripping or adding insulation usually plug the most obvious heat leaks and are, consequently, the least expensive and most effective.

Heating potential. Because each solar greenhouse installation includes many variables, it is impossible to estimate exactly the contribution it can make to heating a home. Designers of new homes that incorporate a solar greenhouse claim it supplies as much as 85 to 90 percent of the heat needed. Where the greenhouse is attached to a south wall of a more conventional home, heat contributed can be as much as 60 percent.

In addition to its heat-producing potential, there are several other reasons why a greenhouse can be a useful addition to a home.

Other benefits. Unlike a furnace or even an active solar heating system, a solar greenhouse can add extra living space to a home. A well-done solar greenhouse can be an attractive gardenlike addition that can directly increase the value of your home. In winter, the plants in the greenhouse help purify and humidify stale, dry air.

For the homeowner who enjoys working with plants and likes the taste of home-grown tomatoes and vegetables, a solar greenhouse makes it possible to produce food crops almost the year around, even in the coldest climates. Not only can you calculate the pay-back period for your investment in a

Plastic film and PVC pipe will produce a functional, low-cost solar greenhouse that can be dismantled during warm seasons.

solar greenhouse by figuring the savings from utility bills, you can deduct any savings you make in visits to the local supermarket. A well-designed solar greenhouse built with quality materials, whatever its energy-saving features, will last as long as the house itself.

Building a solar greenhouse

The builder has many choices to make when planning a solar greenhouse, but regardless of the type of glazing used or the specific materials selected, the basic design remains the same.

What follows are some of the construction details to be considered in preparing final building plans for a substantial structure. While extremely rudimentary structures may heat just as more solid ones do, the latter offer the most complete lessons for would-be designers.

Foundation. Proper support is necessary and is required by local building codes. It can be provided by one of several common constructions.

A poured concrete footing with ½-inch reinforcing bar can be topped with a cement block stem wall or a poured concrete stem wall. A concrete footing, stem wall, and slab floor poured in one monolithic pour is another common approach. (For more information, see the *Sunset* book *Basic Masonry Illustrated*.) In any case, ½-inch anchor bolts should be imbedded in the stem wall to secure the sill plate of the basic wooden frame structure (see drawing on next page).

Whatever the footing planned, it is important to provide proper exterior insulation. One-inch rigid polystyrene sheeting should be placed against the exterior of the foundation extending at least 2 feet

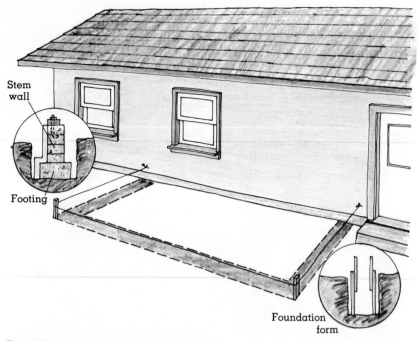

For permanent installation, you'll need continuous foundation with mud sill anchored in place by reinforcing bar. Location must accommodate inlet-outlet vents in house wall.

Angled wall increases heat gain. Framing is simple; be sure to flash roof joint.

below ground level or well below the frost line. This prevents heat inside the greenhouse from being lost to the ground outside.

Walls. The south wall is the primary source of sunlight, and it should be double-glazed with glass, fiberglass, plexiglass, or polyethylene. Since these materials come in standard sizes, you'll save on construction costs if you plan the greenhouse around the sizes most available at your lumberyard.

Traditional design suggests a south wall leaned at 60° for maximum exposure to the winter sun. In climates with mild winters and very hot summers, there are advantages to using a vertical south wall. Vertical walls reduce energy absorbed from October to April by about 25 percent. But they cut heat gain in half in the summer, and they enclose a larger amount of usable space.

As the basic concept is to prevent heat loss in the winter and overheating in the summer,

east and west walls need not always be glazed. If not, they should be solid, well insulated, and caulked.

The rear wall should contribute to heat storage and the efficiency of the greenhouse by having significant mass. In a newly designed home, it can be of masonry and can function as a direct storage mass. Alternatively, drums or bottles of water along the house wall can store heat.

Floors. Concrete, brick, flagstone, or any similar material can provide a solid floor and contribute mass for heat storage. For greatest efficiency, the floor can be underlaid with solid sheets of insulation to prevent heat loss through the ground.

Roof. In designs with an angled south wall, a regular roof can be built from the house to where it joins the wall. Where a vertical south wall is used, the leading edge of the greenhouse roof should be glazed to let in the

maximum amount of winter sun. The solid portions of the roof should be built of standard roofing materials and well insulated. In summer, glazed areas of the roof can be shaded with regular windowshades, fabric, or a material of your choice.

Ventilation. During cooler weather, the hot air that collects under the roof of the greenhouse should be vented directly into the house. Vents to permit cooler air from the house to flow back into the greenhouse should be provided near the floor. Roughly 1 square foot of vent space should be provided for every 25 square feet of south-facing transparent surface.

Air can be circulated with a thermostatically activated fan, or it can be permitted to circulate naturally through the vents just described. A manually controlled roof vent in the greenhouse may be required to dissipate hot air during the summer.

Vents to provide cross ventilation are necessary in the end walls. A vent on what is usually the windward side should be built near the floor. A vent on the side away from wind should be built near the roof. Plan about 1 square foot of vent area for every 6 square feet of floor area. If an exhaust fan is used, the size of the vents can be reduced.

All vents should be designed so they can be closed.

Vents into the living area of the house may have to be closed during winter nights to prevent warm house air from being lost when temperatures in the greenhouse fall, and during hot summer days to keep out excess heat. All vents should seal tightly when closed.

To maintain temperatures in the greenhouse at high enough levels to prevent damage to plants (about 40°F/5°C), heat from the house can be allowed to circulate back into the greenhouse. But it is more common, and more efficient, to provide back-up heat for the greenhouse. Electric baseboard space heaters are most common as they can be activated easily with a thermostat.

In addition, heat loss during prolonged gloomy weather or on cold nights can be curtailed by the use of manually extended shades or shutters. (The same units can cut down excess summer heat.) This system can be effective, but it requires daily attention and may be bothersome as a daily chore.

Heat storage. Mass that absorbs heat and releases it when the atmosphere cools is the basis of the solar greenhouse. Mass can be water, rocks, or almost any masonry material.

The size of storage mass varies quite a bit, depending on the material used. Approximately 2 cubic feet of water or 75 pounds of rock are required for every square foot of glazed area. This is three times as much rock as water by volume and five times as much by weight, which explains why water is considered a desirable medium for heat storage.

Water is commonly stored in 55-gallon drums that have been painted black or some other heat-absorbing color. The drums can be placed where they are struck fully by the winter sun or used in partial shade to support planting beds. Water can also be stored in culverts, plastic bottles, or any one of dozens of other containers.

Rock used for heat storage should be contained in wire mesh or otherwise be well ventilated. Air must be permitted to circulate between the rocks to carry heat.

The more storage, the more heat the greenhouse will retain. It is possible to overdo storage, however; too much storage will warm too slowly, reducing the amount of heat available to the house.

And the last word on heat storage is ingenuity: one experimental solar greenhouse built in Oregon utilizes a concrete water tank that not only absorbs heat, but supports tilapia, a tropical food fish.

Finishing. To make the completed greenhouse as efficient as possible, first add caulking and weatherstripping.

Solid walls should be well insulated and areas around doors and vents sealed.

To prevent phototropism (the tendency for plants to grow in the direction of their primary light source), the upper areas of the rear wall and the area under the roof should be painted with bright white or any other reflective paint.

Stored water should be treated with ¼ cup of sodium dichromate for each 55-gallon drum (or equivalent) as a corrosion inhibitor (unless you plan to raise fish).

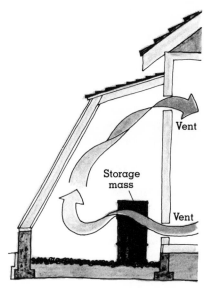

For a solar greenhouse to work well, it must have thermal mass to store heat, and must have vents high and low in the house wall so convection will move warmed air indoors. The vents must close to keep excess heat and chill out.

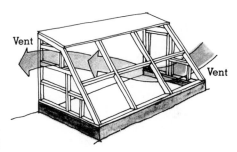

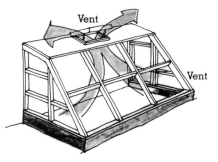

In addition to venting into a house, a greenhouse must have its own vents to reduce heat, humidity. If you cross-ventilate, place one vent low in windward wall, the other high on downwind side. Roof vents may be preferable in regions of high summer heat.

Solar Greenhouses **43**

THE POSSIBILITIES OF PASSIVE SOLAR HEATING

In passive solar space heating, the house stores heat within its own structure for use at night and on cloudy days, and distributes the heat with a minimum of mechanical aid. In short, the house becomes its own heating system.

The main challenge is to get direct sunlight into as many areas of the house as possible in direct-gain designs, or as much sunlight onto a storage mass as possible in indirect-gain designs.

In direct-gain systems, materials with high heat-holding capacity are used in floors and walls to store the heat evenly throughout the building.

In indirect-gain systems, the materials with this high heat-holding capacity may be part of the structure or may be separate units such as water-filled drums.

Once heated by the sun's rays, the masses reradiate stored heat slowly; in indirect-gain designs, air currents (sometimes augmented by fans) are depended upon to distribute the heat to areas away from the storage mass.

In existing buildings, developing efficient storage and distribution can be somewhat difficult compared to collection.

On these pages we show three basic devices for making use of the sun's heat in conventional houses. The *Sunset* book *Homeowner's Guide to Solar Heating & Cooling* offers a detailed survey.

This skylight makes several points about passive solar space heating. Hinged, insulated shutter inside frame closes at night or on cloudy days to stop heat loss from interior; its white canvas face reflects extra radiation downward while it is open. (Surgical tubing along all edges forms a tight seal when shutter is closed.) Also, opaque north wall of skylight minimizes heat loss on shaded side, a problem on sunny but cold days. Design: Valerie Walsh, Green Horizon.

Black tubes filled with heat-holding eutectic salts are in experimental use as a passive device. Tubes are set within a double-glazed south-facing window. An interior plywood panel has low vents for cool-air return, and high, closable ones to allow warmed air back into room, or keep it out, as required. Design: Richard Cushing, Arc-Sun Design.

This adaptation of a Trombe wall (named after inventor Dr. Felix Trombe) is a south-facing transparent cover attached to an existing masonry wall. Wall is painted dark. Simple 2 by 4 frame is attached to wall so it can be removed in summer. Sheet of polyethylene film is attached to each side of 2 by 2 facers, which in turn are tacked to basic frame. Wall must have top and bottom vents to allow warm air to circulate inside, as in situation described above. Design: Gary Socha and Dr. Francis Wessling.

INSULATING THE SHELL

In dealing with air movement and the gain or loss of heat through glass, the whole purpose is to establish flexible control so you can use whatever free heating or cooling nature gives.

But when it comes time to deal with the main structure of the house, flexibility is no longer a factor. Now the goal is to give your heaters and coolers the steadiest, most predictable environment in which to work. This means efficient resistance to heat exchange.

In general, conventional structural materials are no great shakes as insulators. The tables on page 48 point this out all too clearly. For the most part, we think straightaway of fiberglass, cellulose, foam, or some other insulation as the solution. And it is the best start, especially in new construction where you do not have to settle for less than the ideal. In older structures, the finished building sometimes gets in the way of installing certain types of insulation.

In any case, new building or old, there are helpful alternatives or supplements to standard commercial insulation.

In truth, the great insulator is not solid material, but air. The more tiny pockets in which it can be contained, the better the effect. For example, a solid sheet of plastic is not a tenth as powerful an insulator as the same plastic in the form of foam. With this thought in mind, cork paneling may be better than solid wood paneling for surface insulation on a wall.

The other thought that needs to be kept in mind in approaching an existing house is this: sometimes it is more efficient to make a small part of the house very comfortable and live as much as possible in that part, while the rest of the building is allowed to grow a bit too warm or too chill for easy use. Large buildings, and especially rambling ones, usually lend themselves to the creation of a cozy nook at comparatively modest cost and effort.

Heating and Cooling Savings through Insulating

Percentages in the chart assume that the house is weathertight and that windows are well protected.

Action	Range of savings	Your potential savings
Insulation ceiling from none to R-19 (See pages 48–49, 52–55)	Savings range widely, depending on climate. In mild climates, savings on heating can reach 40%; if heavier insulation is required, figure drops slightly. Savings on cooling are about 20%.	
from none to R-30 (See pages 48–49, 52–55)	Can reduce heating bill by as much as 40% in severe winter climates, more in milder climates, though cost may outweigh advantages gained.	
Insulating floors from none to R-11 (See pages 48–49, 62–65)	Typical savings are about 10% of heating bill. In severe winter regions, figure is lower. In mild winter regions, underfloor insulation is ineffective.	
from none to R-19 (See pages 48–49, 62–65)	In severe winter regions, savings on heating range from 10 to 13%. Savings can be slightly more where winters are less severe, though cost is likely to outweigh gain.	
slab floor (See page 63) Insulating walls (conventional 2 by 4 framing) from none to R-11 (See pages 48–49, 56–61)	Two-inch-thick perimeter insulation can trim heater use 5%, sometimes more. In most regions can save 10% only if entire living space is thoroughly insulated.	
from none to R-19 (See pages 48–49, 56–61)	Can save 10% in severe winter regions, more in milder climates, but cost effectiveness is poor.	

Planning to insulate

The owner who wants to make an uninsulated home energy efficient may instinctively plan basic ceiling, wall, and underfloor insulation as the first step. After all, the manufacturers of insulating materials and the power companies have promoted this common approach to home insulation heavily in recent years.

To be sure, it is important. But in considering where to insulate and when, make sure the steps you take fit into an overall plan to improve the efficiency of your home. For example, it may well be desirable to lay sufficient insulation in the ceiling to plug a major source of heat loss. But it may be more important in terms of time and budget to ventilate your attic before taking on the job of insulating a wall.

As a basic rule, you'll want to concentrate on insulating the space that you heat and occupy. As a matter of priority, you'll want to start at the top and work down. Homes that have been built since energy costs began to rise sharply usually have been well insulated at the time of construction. But even some of these have inadequate insulation. And older homes may have insulation that has compressed or settled enough to lose much of its insulating value.

Many older homes, of course, were never insulated at all. Energy was inexpensive when they were built, and there was little incentive to spend money on insulation. Unhappily, many occupants of older uninsulated homes made the decision not to insulate on the grounds of expense. If you have such a home and wish to improve it, read on.

Before running to the local building center, remove electrical outlet covers and peek through to determine which, if

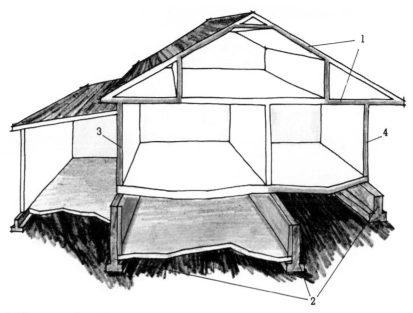

Where to begin

The most effective sequence for insulating an existing building in cold-winter regions is this: 1. the roof above a finished attic or the ceiling below an unfinished (and vented) attic; 2. under floors; 3. walls between heated rooms and unheated spaces like a garage or basement; and 4. exterior walls.

This sequence is supported by studies showing that heat losses in an uninsulated 1,500-square-foot building follow this pattern: 45 percent through the ceiling, 20 percent through the floor, and 12 percent through walls. (The remaining 23 percent is lost through window glass and air infiltration.) The first three in the series also are usually the most accessible places. In mild-winter regions, underfloor insulation drops in importance; see the table of R-factors on page 48.

any, walls have already been insulated. Go to the basement or crawl space to check under floors, and to the attic to check ceiling insulation. Find out whether or not the existing insulation is in serviceable condition.

In establishing a timetable for the insulating work you elect to do, give priority to those areas that are responsible for the greatest heat loss, and begin with the one that is easiest to reach. For example, if you have an unused attic with exposed ceiling joists, you have already found the first target. As the sketch of priorities shows, this is far and away the point of greatest weakness.

As you consider insulating floors or walls, estimate the difficulty of the job and the labor involved. You may be facing a point of diminishing returns, where the cost and effort will never be recovered simply through lower fuel bills. Keep in mind, though, that marginally cost-effective insulation today may prove to be worthwhile over a long period. Energy costs and the price of insulating materials will continue to rise.

Home energy efficiency will be realized through the cumulative effect of completing many individual tasks. It is not necessary to finish the entire, massive job in one exhausting marathon rush. Set a realistic schedule and enjoy the satisfaction to be found in completing each step along the way.

How much insulation?

R-value totals shown below are required in new home construction and may be considered the ideal in adding insulation to attics and under floors in existing buildings.

The figures for walls are for new construction only; no standards have been set for insulating existing walls. Where there are two R-values, the first is recommended for gas or oil heated homes, the second for electric.

The zones are defined by heating degree days, the cumulative total of daily average temperatures below 65°F. The map is only general; consult your city or county building department for accurate local information.

There have been efforts to establish a counterpart system of cooling degree days, to help standardize insulation needs in regions where air conditioning is important. Because insulation is not as efficient in cooling, and because relative humidity plays an important role, standards have been impossible to set to this date.

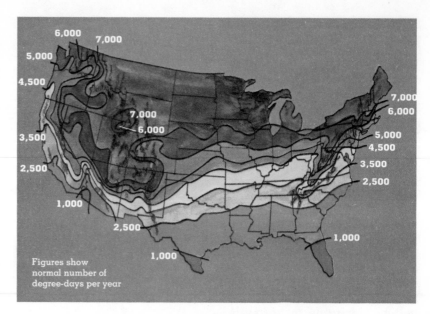

Figures show normal number of degree-days per year

| | FHA-suggested R-values | | |
Degree days	Ceilings	Walls	Floors
7001 or more	38	17	19
6001-7000	30/38	12/17	11/19
4501-6000	30	12/17	11/19
3501-4500	30	12/17	11/19
2501-3500	22/30	11/17	0/11
1001-2500	19/22	11/12	0
1000 or less	19	11	0

R-values of common building materials

Roofing materials	R-values
asbestos cement shingles	0.21
asphalt shingles	0.44
asphalt roll roofing	0.15
½" clay tile	0.05
½" slate	0.05
wood shingles	0.94
wood shakes (1" butt)	1.69
⅜" built-up roofing	0.33

Ceiling and wall coverings	R-values
¾" acoustical tile (mineral)	2.38–2.86
¾" acoustical tile (wood or cane)	1.89
⅜" gypsum board	0.32
½" gypsum board	0.45
¾" cement plaster	0.15
⅝" gypsum plaster	0.39
¼" plywood	0.31
⅜" plywood	0.47
½" plywood	0.62
¾" plywood or wood panels	0.93
$7/16$" medium density hardboard	0.67

Siding and sheathing	R-values
wood shingles (7½" exposure)	0.87
wood ½" by 8" lapped siding	0.79
wood ¾" x 10" lapped siding	1.05
aluminum siding with air spaces	0.61

Siding and sheathing	R-values
⅜" plywood sheathing	0.38
$25/32$" insulating sheathing	2.06
building paper	0.06

Flooring and floor coverings	R-values
⅝" particleboard underlay	0.82
¾" plywood or wood subfloor	0.94
maple, oak and similar hardwood finish flooring; per inch	0.91
fir, pine, and similar softwood finish flooring; per inch	1.25
carpeting on pad	1.23–2.08
½" tile	0.05
linoleum	0.08
felt underlay	0.08

Masonry materials	R-values
stucco; per inch	0.20
common brick; per inch	0.20
face brick; per inch	0.11
4" hollow clay tile	1.11
8" concrete blocks	1.11
with lightweight aggregate	1.50
concrete:	
sand-and-gravel; per inch	0.08
lightweight; per inch	0.19–0.60
pumice aggregate; per inch	1.43

About "R"...

A visit to your utility company or building department is a quick and sure way to find out how many "R"s are considered adequate for where you live. This number, called an *R-value*, is the measure of all insulation.

It's a good idea to understand the system so you can figure out for yourself how much insulation you need. The idea of an R-value for building materials is fairly simple.

Every material conducts heat with greater or lesser ability. Materials that conduct heat well are great for cooking, but rotten for insulation. Resistance to heat transfer makes good insulation.

The arbitrary measure "R" is the ability of 1-inch-thick wood to *resist* heat transfer. Everything else is either a fraction (the R-value of aluminum is 0.0007 per inch), or a multiple (the R-value of expanded polystyrene foam is 4.16 per inch). More graphically, coffee at the boil is hot enough to fry your fingers if you touch an aluminum pot full of it, but the same coffee heats a styrene foam cup barely enough to make the outside warm.

All standard building materials have been tested and assigned R-values. On the facing page we show many of the commonly available materials and their R-values.

Assemble the materials into walls or floors, and the sum of the assembly is slightly greater than the sum of its parts. This is because air insulates; it both fills any gaps and clings to surfaces; and it adds R-value wherever it settles. The sketches show R-values for several uninsulated standard assemblies.

Here the plot begins to thicken.

... and "U"

The greater the difference in temperature between two masses, the faster heat will move toward cold.

Engineers calculate the speed at which heat moves through any given number of "R"s as the "U" factor, and use the latter figure to decide how many "R"s are needed to provide adequate home insulation in differing climates. This is why you can be comfortable in a house in some California beach town, but would freeze to death in the same building if it were in International Falls, Minnesota.

R-values of typical uninsulated assemblies

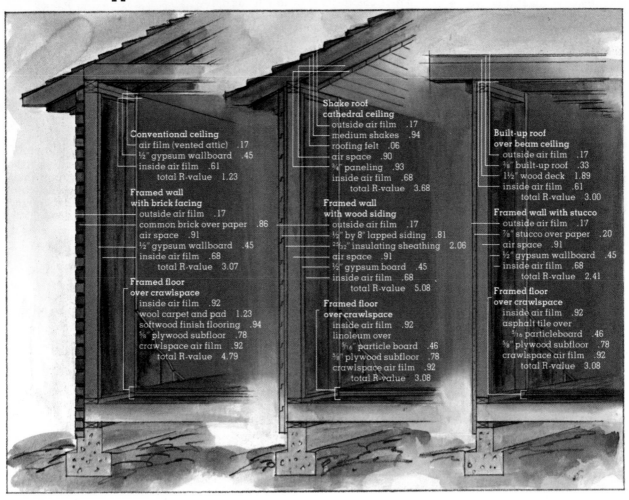

Conventional ceiling
- air film (vented attic) .17
- ½" gypsum wallboard .45
- inside air film .61
        total R-value 1.23

Framed wall
with brick facing
- outside air film .17
- common brick over paper .86
- air space .91
- ½" gypsum wallboard .45
- inside air film .68
        total R-value 3.07

Framed floor
over crawlspace
- inside air film .92
- wool carpet and pad 1.23
- softwood finish flooring .94
- ⅝" plywood subfloor .78
- crawlspace air film .92
        total R-value 4.79

Shake roof
cathedral ceiling
- outside air film .17
- medium shakes .94
- roofing felt .06
- air space .90
- ¾" paneling .93
- inside air film .68
        total R-value 3.68

Framed wall
with wood siding
- outside air film .17
- ½" by 8" lapped siding .81
- 25/32" insulating sheathing 2.06
- air space .91
- ½" gypsum board .45
- inside air film .68
        total R-value 5.08

Framed floor
over crawlspace
- inside air film .92
- linoleum over
- 5/16" particle board .46
- ⅝" plywood subfloor .78
- crawlspace air film .92
        total R-value 3.08

Built-up roof
over beam ceiling
- outside air film .17
- ⅜" built-up roof .33
- 1½" wood deck 1.89
- inside air film .61
        total R-value 3.00

Framed wall with stucco
- outside air film .17
- ⅞" stucco over paper .20
- air space .91
- ½" gypsum wallboard .45
- inside air film .68
        total R-value 2.41

Framed floor
over crawlspace
- inside air film .92
- asphalt tile over
- 5/16" particleboard .46
- ⅝" plywood subfloor .78
- crawlspace air film .92
        total R-value 3.08

Insulation materials

Nine different materials have been in regular use as insulation in recent years. The most familiar of them are the ones do-it-yourselfers can install readily; especially included among these are fiberglass and rock wool in the forms of blankets or batts.

Of the other materials, some can be installed by do-it-yourselfers; a few can be installed only by professional contractors with elaborate equipment. These most notably include the foams.

Brief descriptions of each of the materials follow. The forms the materials take are described in adjoining columns.

Vermiculite. Packaged for loose fill, vermiculite is made from expanded mica. It can be blown or hand poured, and is mainly used in attics and oddly shaped cavities. It tends to absorb moisture and become mushy, which diminishes its already low insulating value of R-2.1 per inch of thickness. It is inorganic and will not burn.

Perlite. This volcanic rock has a slightly higher R-2.7 per inch insulating value, but is similar to vermiculite in all other characteristics.

Rock wool. Along with fiberglass, rock wool is the most common home insulator. It is spun from slag rock, then formed into blankets and batts, or shredded for hand pouring or blowing. Per inch of thickness, it is rated R-3.7 in blanket or batt form and R-2.9 as loose fill. Rock wool is inorganic and will not burn.

However, vapor barriers commonly bonded to batts and blankets are flammable. Usually they are not exposed. In cases where the vapor barrier is left open, it is required by most building codes to be covered with ½-inch gypsum board or similarly fire-resistant paneling.

Fiberglass. It is usually similar in use and form to rock wool, but also comes in rigid boards. Per inch of thickness, its R-values are slightly lower than rock wool (R-3.1 for blankets or batts, R-2.2 for loose fill, and R-4.5 for rigid boards). Fiberglass insulation will not burn, but its vapor barrier will; it should be handled as rock wool.

Polystyrene. These rigid plastic boards dent easily and are highly combustible. But they are useful for below-grade and some exterior wall applications. R-value per inch of thickness is 3.5 to 5.4, depending on density.

Cellulose. This is shredded insulation made from recycled paper products. It is normally machine blown into covered walls, or machine blown or hand spread between joists in attics. It is quite dusty. Per inch of thickness, it is rated R-3.6.

Only use cellulose that has been treated with fire retardant and made rodent resistant. Because cellulose is flammable, it should not be installed within 3 inches of any junction boxes, recessed light fixtures, or other electrical connections.

Urea formaldehyde. Used as a foam to fill wall cavities, it is relatively expensive and requires a skilled contractor to install. It has a high insulating value of R-4.2 per inch of thickness.

Though it is fire resistant, it will burn in intense heat. To protect it from fire and from weather deterioration, urea formaldehyde should not be left exposed. It may shrink slightly and is water retentive. While curing, it releases formaldehyde gas, so all interior areas should be well ventilated. While this book has been in process, some manufacturers have withdrawn this material for further testing.

Urethane. Either as foam or as rigid boards, urethane has a high R-4.5 insulating value. But like urea formaldehyde, if used as foam, it requires a skilled contractor for installation. Urethane is flammable and recommended only for roofs and other exterior surfaces. It releases cyanide gas when burned.

All foams continue to be tested and evaluated for their release of toxic gases.

Isocyanurate. These rigid boards, rated R-9 per inch of thickness, are the most effective insulation available, but also the most expensive. The material is moisture resistant. Like urethane, it gives off cyanide gas when burned.

Forms of insulation

The materials used in insulation come in five basic forms adapted to various uses.

Loose fill. It can be blown into finished walls (though it tends to settle and lose effectiveness with time), or it can be blown or hand-poured between joists in attics or into odd-shaped, hard-to-reach cavities. Vapor barrier must be provided separately.

Blankets. These large rolls of insulation are particularly easy to use during new home construction or in homes with exposed framing in the attic or basement. They come in varying thicknesses and widths to fit standard construction spacing. Blankets come with or without vapor barriers (kraft paper or foil). Long lengths can be unwieldy to fit around pipes, wiring, cross-bracing, or the like, but can be cut to fit.

Batts. Precut lengths of 4 and 8 feet make batts easier to use than blankets, if you are working in these modules (for example, in a room with an 8-foot ceiling). They are somewhat less flexible than blankets, so less convenient for filling long runs between exposed framing, especially in cramped attics or crawl spaces.

Rigid boards. These precut slabs are most commonly used in new construction, but can be used when insulating between exposed ceiling beams, when adding an insulating wall in a basement, or when insulating the perimeter of a slab floor. They come in varying thicknesses and a variety of standard sizes, including 4 by 4 feet, 4 by 8 feet, and 2 by 8 feet. Some are unbacked; others are backed with foil, kraft paper, or building felt.

Foams. Only experienced contractors can apply foams. They can be blown into closed walls or onto flat or gently sloping roofs.

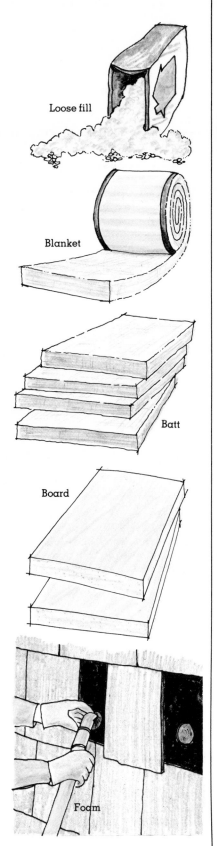

Loose fill

Blanket

Batt

Board

Foam

Vapor barriers

The R-value of most vapor barriers is 0.00, but they remain an important part of insulation.

When heat moves to cold, it carries vapor in the air along with it. In the winter months, everyday activities like showering, cooking, washing, and running a dryer can put from 5 to 30 pounds of water into the air each day. To keep this moisture from penetrating into floors, walls, or roofs, it is necessary in all but the driest climates to install vapor barriers that repel moist air before it reaches colder surfaces.

Moisture that collects on the inside of cold exterior wall surfaces can soak in and cause outside paint to blister. It can accumulate enough to stain inside walls and cause rot in the house's structure. When it collects in insulation, the dampness diminishes the effectiveness of the insulation.

Vapor barriers aren't a substitute for good ventilation. A properly ventilated house and attic will eliminate a good portion of the moisture that builds up in a home. But vapor barriers do protect the effectiveness of insulation.

Types of vapor barriers. Blanket insulation is available with kraft paper or foil vapor barriers. These are usually the easiest to install.

If you are insulating with loose fill, blankets, or batts with no vapor barrier, a separate barrier of polyethylene (minimum thickness 2 mils, or $2/1000$ inch), laminated asphalt-covered building paper, or foil-backed gypsum wallboard (for walls and ceilings only) will provide a good substitute.

Where vapor barriers go. Generally, place the vapor barrier toward the surface that is warmest in winter. In the attic roof, insulation between the rafters should have the vapor barrier face down. In unfinished attics, insulation placed between floor joists should have the vapor barrier facing toward the living area below. The vapor barrier faces in when used in exterior walls and up when used between joists under the floors of heated rooms.

The ground under a house should be covered with 4 to 6-mil polyethylene sheeting to prevent moisture from being drawn out of the earth.

In an excessively damp climate, as in some areas of the South, the exterior air may normally carry more moisture than the air inside a house. In that case, vapor barriers are placed toward the dampest air and not necessarily toward the living area. If this is a consideration where you live, consult your local utility company or building material supplier for advice.

Vapor barrier alternatives. Foam insulation (applied wet or in sheets) creates its own vapor barrier. In cases where loose fill is blown into walls covered on both sides, the installation of small louvered vents can help solve the moisture problem. Local building codes usually give specific requirements.

One common vapor barrier alternative is to paint interior wall surfaces with two coats of a paint that has a "perm rating" (permeability) of less than 1. If that is unavailable, two coats of high-gloss enamel or a varnish-based, pigmented wall sealer will offer a reasonable substitute.

If both interior and exterior wall surfaces are sealed, they should have vents installed to prevent moisture from being trapped and accumulating in the wall. These simple metal vents, resembling round cooky cutters, fit into holes cut in the exterior siding.

Insulating cathedral ceilings

Cathedral ceilings with slopes and exposed beams are attractive, but adding insulation to them poses some considerable problems.

Unhappily, it is not easy to achieve the most desirable R-value (R-30 in most climates) in a poorly insulated roof, unless you do so from the outside when you replace the roof. Rigid foam insulation sandwiched between layers of felt directly over the roof decking should give the R-19 level.

If you must insulate from the inside only, you can usually add R-9 in the form of 2-inch-thick panels of vinyl-faced pressed fiberglass, without hiding the beams from view. The panels can be cut to fit snugly against the ceiling between beams, and they're light enough to be secured with quarter-round molding nailed to the sides of the beams.

The standard panels are 4 by 16 feet. They come with a variety of textured vinyl faces and can be cut easily with a knife and straightedge.

If the heating bill is eating you alive primarily because of your cathedral ceiling, and if you're willing to sacrifice the appearance of the exposed beams, you can use them as the anchor points for ¾-inch tongue-in-groove board; fill the cavities thus created with as much insulation value as you choose. You can also use gypsum board in place of the ¾-inch board, but its weight may require handling by a professional installer.

Vinyl-faced insulating board comes in widths to fit most standard beam spacing. One advantage: it can be held in place by light strips of molding tacked to beams, permitting later removal without damage to original ceiling's appearance.

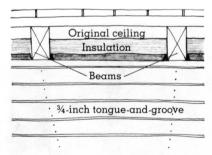

Cross section shows how insulating board is secured in place by molding tacked to sides of beams.

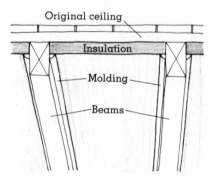

Permanent alternative to rigid board insulation is use of ¾-inch tongue-in-groove boards nailed to faces of beams; batt or blanket insulation then fits into cavities created by new ceiling surface.

Lowering ceilings

Glamorous as old-fashioned 14-footers and cathedral ceilings are, lowering them to 8 feet has considerable potential for saving energy.

Reducing the volume of air your heaters must warm is a direct way to improve their relative capacity. They work less because they have less space to heat.

However, the idea needs to be approached with caution. First, lowering a ceiling often is a major undertaking. Second, it does little good unless the new ceiling can be insulated enough to slow heat transfer effectively. R-11 should be considered minimum.

A great majority of ceilings lowered these days use acoustical tile panels set into prefabricated metal channels suspended from ceiling joists on wires. There is little or no cutting; the skill is in establishing and holding a level.

Most systems use a 2 by 4-foot grid. All will bear the added weight of blanket or batt insulation, which can be installed as you place the tiles.

The limitation, especially in fine old houses, is design compatibility. These systems are relentlessly modular, and thus modern in appearance.

The alternative is a regular framed ceiling, starting with 2 by 4s secured to wall studs on opposite sides of the room. Joists are toe-nailed to them, and usually covered with gypsum board; you insulate as you go. It's a job best left to carpenters.

As with many of the projects calling for insulation in an existing building, fitting one room as a snuggery is more likely to pay dividends than doing a whole house because it produces comfort at a reasonable cost while allowing the thermostat to be turned down.

Acoustical tile ceilings fit well into modern decors and are relatively easy for do-it-yourselfers to install; prefabricated metal channels hang on simple wires wound through screw eyes.

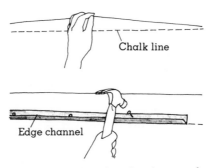

Establish ceiling height, then mark level line on walls with chalk line. Nail edge channels in place.

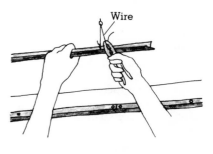

Tie wires to screw eyes driven into original ceiling joists in grid determined by size of panels.

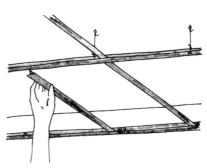

Wire cross channels in place, using string tied between edge channels to establish correct height.

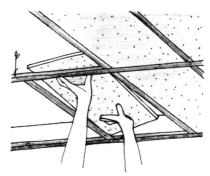

Place tiles in channels, laying insulation atop each row as it is finished.

Insulating attics

More heat escapes through an uninsulated attic than through any other part of a house. In a typical 1,500-square-foot uninsulated house, about 45 percent of heat lost literally goes right through the roof.

You will earn the greatest return for the least investment by beginning your insulating in such an attic. If you're lucky enough to have an unfinished attic with exposed joists and rafters, the job should go quickly and be relative cheap. If your attic is properly ventilated (see "Attic ventilation," page 17), it should only be necessary to lay insulation between floor joists.

On the other hand, if your attic is finished and the interior surfaces covered with wall board or paneling, adding insulation can be complicated and possibly quite expensive. You may need professional help.

You may want to take temporary measures, like using rugs on the attic floor, and delay a thorough insulating job until it can be incorporated into a general attic remodeling project. In any case, make certain a finished attic is properly caulked and weatherstripped early in any effort to improve a house's energy efficiency.

Because most homeowners

Preparatory steps

1. Have at least two work lights (mechanics' trouble lights or similar types that can be moved easily) and enough planking so you can move freely without risk of putting a foot through your ceiling.

2. Build sheet metal barriers on all sides of flues; allow at least 3" on all sides. Fill space with cinders, pumice, or other lightweight inorganic material to keep flammable materials away from flues.

3. Place wood baffles at least 3" from each electrical connection, and between insulation and soffit vent (see drawing): the flammable vapor barriers must be kept away from electrical connections.

Insulating with loose fill

4. Secure 2-mil polyethylene or foil with staples to form vapor barrier for loose fills.

5. After fill is blown or poured, spread and fluff it with a rake. Even out high or low spots.

6. Use board to level fill precisely; notch board if level is below top of joists. Don't compress.

will be insulating unfinished attics, the illustrations on these pages detail the work the average homeowner can complete without professional help.

As a first step in choosing your insulation, find out the recommended R-value for ceilings in your area by phoning the city or county building department.

The depth of the cavities between your joists may limit your options. For example, cellulose — the most effective of the loose fills — must be 8 inches thick to reach an R-value of 30, so would more than cover typical 2 by 6 joists. If you need to attain R-30, batts or blankets with their firm shape might be much easier to use.

A rough guideline to R-values is on page 48; your local building department will give more specific recommendations.

Even if structural limitations do not allow you to insulate to the highest recommended value, some insulation is better than none.

Besides R-value, another factor in choosing materials is your personal response to the irritants in most insulation. Loose fills, especially cellulose, can be dusty. Fiberglass tends to irritate the skin of many people. Because of these factors, maximum ventilation of the attic during work is of prime importance, especially if loose fill is being blown into places.

In addition, wearing goggles, a dust mask, and clothing that protects as much skin as possible will help.

Insulating with batts or blankets

4. Cut batts or blankets with knife; using a board to compress material and guide cutting.

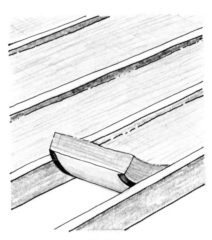

5. Face vapor barrier toward living area except in hot, humid climates (see "Vapor barriers," page 51).

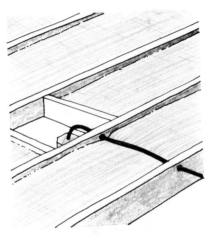

6. Slip batts under wiring, or, where compression would be too great, cut insulation to fit.

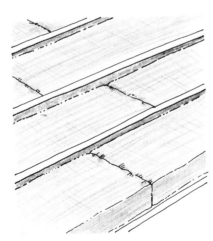

7. Where two ends must butt, cut lengths slightly long so joint compresses.

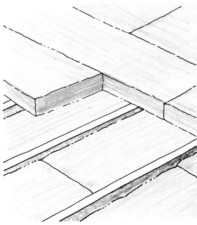

8. If adding to old insulation, use batts or blankets without vapor barriers. Lay across joists.

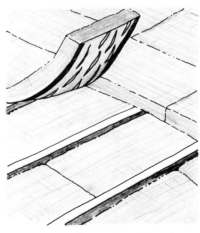

9. If new insulation has vapor barrier, slash thoroughly; then lay face down and across joists.

Covering existing walls

Hardly ever does paneling an existing wall, or hanging a rug on it, allow you to turn your thermostat down any farther than it already is. But either approach can let you live in a good deal greater comfort without turning your thermostat up.

As in zone heating and several other energy-saving plans, the goal should be to create a cozy nook, not to revamp the biggest room in the house.

Look first for a windowless north wall as the candidate for covering. It will be the coldest surface, because it lacks sun warmth all day.

Stucco or other masonry walls are prime targets because their poor R-factor allows them to grow coldest. This, coupled with their smooth surface, allows them to absorb both furnace and body heat most rapidly of all interior wall materials except glass. How-ever, any wall that feels chilly to the touch on cold days is a candidate for covering.

In general, the wall will pay the greatest dividends if it is combined with other heat-saving measures, including thermal shutters for windows and a wood stove or other room heater. On the face of the matter, it might seem as if insulating one of two exterior walls would improve a room's ability to hold heat by half, but, alas, such is not the case. A major gap in insulation is about the same as a weak link in a chain; it is this unhappy fact that has led to the repeated recommendation to create one cozy nook in an existing house rather than attempting to insulate the whole building.

The wall itself contains the elements of the same problem.

Whether you choose an insulative paneling of some substance or a rug depends greatly on how severe the heat-loss problem is. Either solution lends itself to installation by apartment dwellers or other renters, for both coverings can be removable.

As described below, wall hangings are a fairly ready-made project. Paneling over an existing wall offers more choices and more complications.

The project shown on the facing page uses a sandwich of materials fitted flush against an existing wall. Another approach is to place light furring strips on the wall and cover them with insulating board. With heavier furring strips, there is enough depth to install batt or blanket insulation.

Remember that if you use either rigid board insulation or batt insulation with a vapor barrier, the covering must be flame-retardant to meet fire codes. Check with your local building department before you buy your materials.

In any of these approaches, the trick is to secure the materials so they can be dismantled, leaving the wall undamaged.

Wall hangings

Wall hangings should cover as much of the wall area as possible. (Furniture can also help insulate people from a cold wall.) If you have a choice of hangings, for maximum R-value choose wool over synthetics, and shag over smooth-finished textiles. Rough textures keep air stiller along the wall, and air adds insulation.

If the hanging lends itself, stretch it on a frame to add an air space between it and the wall. An air space of at least 1 inch adds 0.92-R to the total. If you do this, the frame must have a continuous edge that fits flat against the wall.

Typically, frames are made of 1 by 2 or similar light wood, with mitered corners reinforced by metal straps. The frames can hang on protruding molly bolts, or, if there is ceiling molding, they can hang on hooks and picture frame wire to spare a rented wall.

A bulletin-board wall

The idea for paneling this room as a temporary measure came about because it was to be used only for a short time as a teenager's bedroom, and the original hand-plastered walls were too fine to be sacrificed to the need for poster space.

The basic notion lends itself to repetition—or variation—wherever renters or apartment dwellers are faced with a dilemma in improving insulation because they are not free to make structural changes. In fact, the idea is useful in making any single room snug without damaging existing walls.

The notion is simple: attach rigid board to the wall with a minimum of screws, then glue an attractive cover over the base. The original combination of ¾-inch cork sheets over ¼-inch hardboard is fairly insulative and not very expensive. Other variations could provide higher R-values, but probably at greater cost.

In this project, the base hardboard is trimmed 4 by 8-foot sheets, placed with the textured sides facing out.

For eventual easy removal, the hardboard was attached with ⅝-inch drywall screws (which can be hard to buy in small quantities; Phillips flathead self-tapping screws are an adequate substitute). Eight to ten screws are required for each sheet of hardboard.

The dark cork (sometimes called burnt cork) sheets are 2 by 3 by ¾-inch thick. In this design the builder used a combination blade on a radial-arm saw to cut each sheet of cork in half lengthwise, to provide foot-wide courses.

Starting with the top course he glued the cork sheets to the hardboard, using acoustical adhesive applied with a small trowel. To prevent a lineup of vertical joints, he began and ended every second row with a 1½-foot-long piece.

Small gaps or holes along the joints are filled with small pieces glued in place.

When the time comes to dismantle this design, the cork will have to be sacrificed. After the hardboard is removed, the screw holes are to be filled with patching plaster.

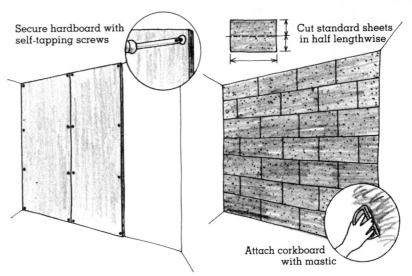

Secure hardboard with self-tapping screws

Underlayer of hardboard attaches to wall with self-tapping screws.

Cut standard sheets in half lengthwise

Attach corkboard with mastic

Top layer of cork is glued in place in pattern of designer's choosing.

Insulating walls

If adequate insulation wasn't installed while the house was being built, making up for the oversight can be difficult and expensive.

Insulating finished walls is a job best handled by a contractor with experience and proper equipment. But a knowledge of the materials and the techniques required will help you in discussing the job with a professional and, perhaps, in finding the right contractor for the job.

However, if you have exposed, uninsulated walls in your house, the work is not complicated. Most often, these will be walls between living space and an attached garage or utility room, or in a basement. (If you have unfinished masonry walls in a basement, see page 64).

Unfinished walls are most often insulated with standard rock wool or fiberglass blankets or batts. In considering your options, compare rock wool or fiberglass with rigid board insulation, which is usually less expensive. In a seldom used area, rigid board might be sufficient.

Because vapor barriers can be flammable, some local building codes require that they be covered with gypsum wallboard or some other fire-retardant surface.

If you select an insulation with a kraft paper or foil vapor barrier, the barrier should, with few exceptions, face the wall that is warm in winter (see "Vapor barriers," page 51).

Take the same precautions with wall insulation as you would with attic insulation: keep it at least 3 inches away from lights, heat ducts, chimneys, and any heat-producing equipment, and don't obstruct the flow of air from natural air vents.

If vapor barrier faces out, staple its flanges to sides of studs at both edges of cavity.

If vapor barrier faces in, staple one flange to side of stud, then fold insulation into cavity.

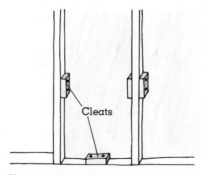

Use scrap wood cleats to hold insulation board in place.

Insulating finished walls

There are several ways to insulate walls that already have finished interior and exterior surfaces. The processes are complicated and currently not cost-effective.

If you've already invested the materials and time to insulate your attic and under floors, and to cut off the more obvious heat leaks around doors and windows with caulking and weatherstripping, you have probably cut heat loss by more than 85 percent.

In a typical 1,500-square-foot home, heat loss directly through uninsulated walls is estimated at about 12 percent. No homeowner should be discouraged from doing a thorough job of insulating, but it would be imprudent to begin to insulate walls without knowing how complex the process may be.

Insulation used in closed walls is either a loose fill, like cellulose fiber, or a chemical foam, like urea formaldehyde. Regardless of the material selected, wall cavities between studs and above and below windows and fire stops have to be filled by insulation blown in through holes bored either in the exterior or interior walls.

As holes must be bored every 1½ to 2 feet all around the house, and probably at various levels, the average home may require the boring of several hundred holes. And each hole has to be sealed, and the wall surface refinished, after the insulation is in place.

In rare cases, the job is made easier if cavities between studs can be reached through an attic, and if there are not too many fire stops or other obstructions in the walls. Loose fill can then be poured directly down into the wall.

Unless a house has a masonry exterior or metal siding, holes are normally bored in the outside walls. Clapboard siding or shingles can usually be pried loose and replaced after the first insulation has been added.

The first holes are drilled high up on the walls, and a plumb bob is dropped through the hole to check for fire stops, braces, or other obstructions. Every time an obstruction is discovered that could restrict the spread of the insulating materials, a new hole must be drilled below the obstruction.

If loose fill insulation is used, it is likely to present some difficulties. First, loose fill tends to catch on nails, pipes, wiring, and other obstructions in the wall and doesn't spread evenly. Second, over a period of time loose fill insulation will settle and lose some of its insulating value, leaving portions of the wall with no insulation at all.

Foams are more successful in filling wall cavities. But they are more expensive than loose fill, and they have properties that many homeowners are not comfortable with. The foams will burn under intense heat and give off toxic gases. Urea formaldehyde has been the most common foam during the past several years, but the fact that it gives off formaldehyde while it cures has caused the material to lose favor to the point that, as this book goes to press, some manufacturers are no longer offering it.

Most building codes consider the foams reasonable materials to use and their risks insignificant. But the foams continue to be studied both for their flammability and toxicity. Obviously, if urea formaldehyde is used, the house should be particularly well ventilated during the curing process. Failure to do so has led to illness in some susceptible individuals, which in turn has led to mistrust of the product.

Two other negative qualities

This pattern of drilled holes is typically required when insulation is to be blown into walls of an existing building. Most are drilled from outside, but in masonry buildings they may have to be drilled in interior wall coverings. Holes are plugged afterwards.

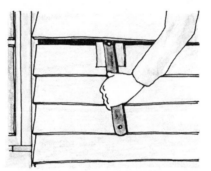

Contractors prefer to pry siding off, then replace it at end of job; it's the only way to hide plugs.

Special rotary saw blades cut holes into which foam nozzles fit, assuring plugs will fit later.

to consider are the general tendency of foams to shrink and lose some of their insulating value, and the fact that some of them carry a high moisture content. Tests of urea formaldehyde show it to contain as much as 1½ pints of water per square foot at 3½ inches thickness, or about 4 pints of water per cavity in a stud wall. Foam-filled walls should, therefore, be ventilated.

On the positive side, the foams are rated highly for their insulating properties — about R-15 in a typical 3½-inch frame wall. Cellulose fiber is rated R-13 for 3½ inches. No other loose fill (fiberglass, vermiculite, or rock wool) can even reach R-11 for 3½ inches.

Because the foams have the potential to be the most effective of the insulations for filling finished walls, considerable effort is being made to eliminate drawbacks like shrinking. And as the cost of energy continues to rise, the high cost of foam may not deter homeowners who wish to improve insulation. Tax credits, granted by many states and by the federal government, are also encouraging the use of materials that might previously have been considered too expensive.

Cross-wall closet

Cross-wall closets require too much material and labor to be money-saving ways toward greater home energy efficiency. But for anyone thinking to remodel a room to give it a new role, at least such a closet may be able to contribute both to the new purpose and to a comfort that does not depend on the thermostat being set at 72°F.

The closet can hold files, tools, or hobby equipment as easily as it holds clothes.

Structurally, the key is the framing of the front wall (a sample design is shown below the color drawing), but as far as energy efficiency is concerned, the heart of the construction is in the back wall. There must be an insulative layer at the rear for the closet to be of much help in keeping a room warm.

Unless you have ample space to devote to the closet, adding that insulative layer can be difficult. Any insulation with a vapor barrier must, by fire code requirement, be covered by a fire-resistant paneling; gypsum board is the preferred choice. All plastic rigid boards also must be covered. And most insulation without a vapor barrier is likely to shed too much of itself into the closet to be welcome.

For these reasons, the most plausible solution is to secure polystyrene or urethane rigid boards to the existing wall with screws, and then to face them with paneling glued to the insulation. (See page 57.)

As in other efforts of the type, it is best to choose a cold, north wall for the closet. It will serve best if the room also is shuttered (or otherwise has its windows covered well) and has a space heater of some kind that keeps it cozy while a lowered thermostat allows the house to cool.

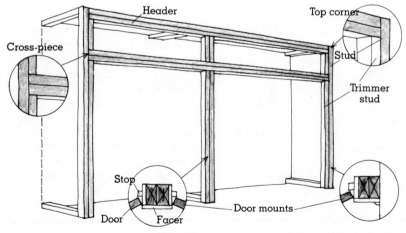

Header
Top corner
Cross-piece
Stud
Trimmer stud
Stop
Door
Facer
Door mounts

A whole-wall sunshade

This solution to a baking-hot bedroom and kitchen suggests itself to people in mild winter parts of the sun belt, where many houses are poorly insulated. It may be a good deal less expensive than opening a west wall to install conventional insulation.

Where winter sun will repay maximum effort to gain heat through windows, much of the warmth lost by shading the wall can be regained. But, in regions with cold, cloudy winters, summertime benefits can be more than offset by the wall's cooling effect.

The wall is very light, and easy for an amateur builder to install. Its framework is ¾-inch thinwall conduit assembled into ¾-inch thinwall pipe fittings. The conduit fits readily into the fittings; each joint is secured by an aluminum nail inserted through drilled holes. Some fittings come with set screws. The job requires caps, elbows, regular tees, and four-way tees, as shown in the drawings.

The surface is 4-inch redwood benderboard air-dried then treated with preservative. The benderboards are woven on alternate sides of the vertical conduits.

Spacing of conduits is governed by the size of the wall. In this case the verticals are 2 feet apart, the horizontals 4 feet apart in a 16 by 20 foot rectangle.

Window openings are dressed by 2 by 3s used as borders.

The surface is about a foot from the original wall, a distance governed by the eave line.

For the reverse solution—turning a wall into a heater—see pages 44 and 76.

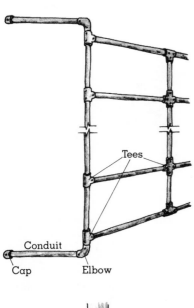

Tees

Conduit

Cap   Elbow

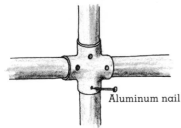

Aluminum nail

Cap   Conduit

Alternate weave of benderboards

Underfloor insulation

The importance of insulating under floors varies greatly, depending on how your house sits on the ground. If your home has no basement and sits above ground with crawl space under the house, the floor could be a major source of heat loss. If your home has a heated basement or sits over a fully enclosed crawl space, insulating is less critical.

In general, homes with heated basements or enclosed crawl space (typical of masonry construction) are best protected by insulating the foundation walls of the house rather than under the floor. Homes with unheated space below the floor are best protected by insulation directly under the floor. Some hillside or split-level homes may have a combination of unheated crawl space and heated basement or garage areas. Each space, of course, is treated separately.

Unenclosed crawl spaces usually have exposed floor joists. They are relatively easy to insulate, even if crawling about under the house isn't the most pleasant way to spend a weekend. Enclosed crawl spaces also have exposed joists, but they offer the material-saving option of in-

sulating the foundation, as shown on the facing page.

Joist cavity installations offer better protection.

Use standard rock wool, or fiberglass blankets or batts, placing the vapor barrier up toward the floor. (Protection from ground moisture is also necessary and will be discussed.) To keep the insulation from sagging, use chicken wire, pressure rods, or any system of wire netting created for the purpose. In frame construction, don't overlook the areas between the top of the foundation and the subfloor.

Once the floor has been insulated and work under the house has been finished, a moisture barrier (4 to 6-mil polyethylene sheeting) should be laid directly on the ground. Sheets should overlap about 6 inches and extend up the exterior walls a few inches; they can be secured with bricks or rocks.

Finally, plug major gaps in the skirt around the house (the covering wall between foundation and floor) to keep the wind from whistling through. If possible, install air vents that can be opened wide in summer and closed to reduce air flow in winter. Some air circulation is necessary, even during the cold months. In regions subject to termites, building codes require vents.

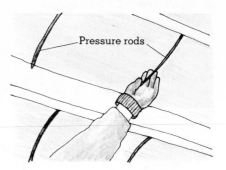

Quick, easy way to hold batt or blanket insulation in place is spring metal rods. Rods are slightly longer than width of joist cavity, so pressure holds them in place. Must be set 2 to 3 feet apart to avoid sags in blankets.

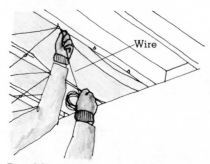

Durable system uses common galvanized wire secured to bottom edges of joists at 2 to 3-foot intervals. A nuisance to install, involving repeated back-and-forth crawling and much overhead work; reliable, though.

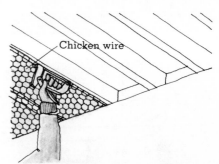

Chicken wire does best job of all at holding batt insulation in place, but installation is hellish task in constricted crawl spaces. Do not use where it interferes with access to plumbing or wiring. Costly.

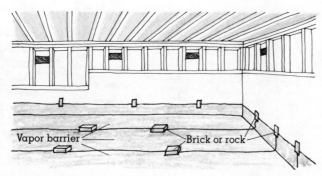

In addition to vapor barrier on insulation, lay polyethylene sheeting on ground, overlapping edges by about 6 inches and anchoring with brick or rock; increases insulation's efficiency.

Foundation insulation

Where there is an enclosed foundation, the technique shown below insulates as effectively as an underfloor installation, but at considerably less material expense, since batts don't have to run the full length of every joist cavity.

The system doesn't work well where building codes require substantial ventilation of a crawl space.

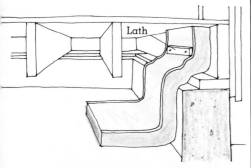

Where you are working parallel to joists, use lath to secure insulation to wall above foundation.

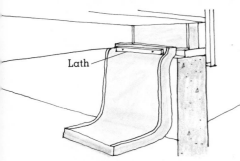

Where you are working between joists, secure insulation to mud sill so edges meet.

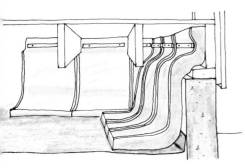

To finish the job, lay a polyethylene vapor barrier on the ground, as shown on facing page.

Insulating a slab

Efforts to insulate under a concrete slab would, of course, be costly and impractical. Still, it is not necessary to live with cold floors as an unavoidable condition of living in a slab-floor home.

To some degree, the selection of interior floor coverings can have a significant effect on the comfort of the living area. Wool rugs on top of rug pads will offer greater insulation than simple vinyl floor covering applied directly to the slab.

It's possible to reduce the amount of cold that seeps in from outside. The foundation around the slab can be protected with rigid panels of urethane, polystyrene, or polyisocyanurate.

Dig the dirt away from the foundation to a depth of at least 2 feet (or below the frost line). Brush the dirt off the foundation and attach the rigid board with mastic. Your building material supplier should be able to recommend the right form of this adhesive. Note the directions for its use: many mastics can be applied only on warm days.

Once the board has been attached, replace the dirt. The dirt will help hold the board in place and will eliminate the need for excessive amounts of mastic.

1.

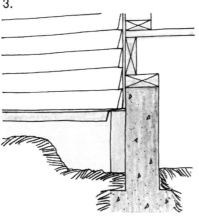

After digging earth away to required depth, brush foundation clean with wire brush.

2.

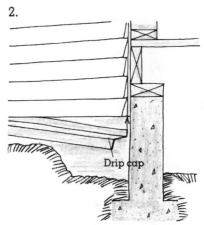

Install metal drip cap; force it under siding if possible, or nail in place and caulk top edge.

3.

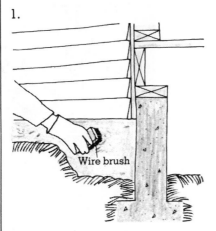

Coat foundation with mastic and press insulation board into place for firm bond.

4.

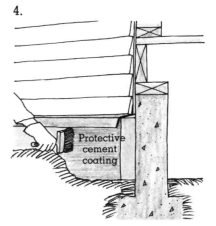

To avoid damage by weather, coat with cement any portion of board that will remain exposed.

Basement insulation

Basement insulation can kill two birds with one stone. It eases the burden of heating the living spaces above, and is the first step toward creating a basement living space, for recreational or other year-round use.

Basements with exposed framing are insulated like any wall or floor with open studs or joists. Basements with masonry walls require framing with either 2 by 2-inch furring strips or standard 2 by 4s, depending on the type and thickness of the insulation chosen. Furring or framing makes it possible to secure gypsum wallboard or other paneling to complete the project.

If you've been considering turning your basement into a recreation room or additional living space, this would be an excellent time to do the job, incorporating it with your insulating work. Framing involves relatively simple carpentry. Just make certain the finished work lies in a perfect plane. Framing on uneven or rough walls may require shims. The goal is to prepare a flat surface for wallboard or paneling.

Either standard rock wool or fiberglass insulation may be used within the framing. If the basement is standard room height, insulation in the form of batts, rather than rolls, may be easier to use when the insulation is to reach from floor to ceiling. This is not always necessary; to be effective, insulation need only reach 2 feet below grade level in mild winter climates, or that far below the frost line in regions of harsh winters. You can save a fair amount of money by stopping the insulation short.

In planning the framing, space the studs to accommodate standard insulation widths. As the framing is of no

Basement hobby rooms can not only serve their own purpose, but can help warm a room above.

structural value, it can be spaced as widely as possible to save time and materials. Standard insulation is produced to be used between studs measured from center to center.

Rigid foam insulation comes cut in standard 2-foot and 4-foot widths, so studs should be spaced to accommodate these widths (inset between studs, rather than measured center to center). Local building codes will probably require both insulating blankets and rigid foam boards to be covered by gypsum board or some other fire-retardant covering.

In a basement partially at grade level, or near it, you probably will want to install an insulating floor in any room you

build. In a basement well below grade, an insulating floor is of less concern, though you may want a wood floor for esthetic or other non-energy-related reasons.

A floor partially at grade is best insulated by the assembly of rigid insulating board, furring strips, and plywood finish floor shown on the facing page. Such a floor is its own vapor barrier as well as good insulation against heat loss to frozen ground.

In a below-grade basement, a simple vapor barrier of the type shown on the facing page is enough, since damp rather than cold is the problem.

The floor goes in first, if you elect to have one.

Basement room is, in essence, simply a box built to fit a space; construction techniques are simple.

Plastic film sandwiched between paired 2 by 4 sleepers is serviceable below-grade vapor barrier.

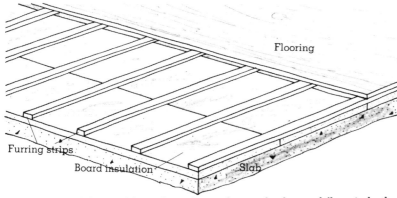

For at-grade slab, rigid board insulation beneath plywood floor is both insulation and vapor barrier.

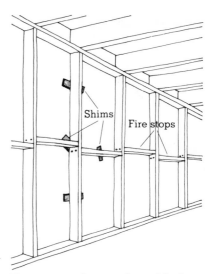

Framing can be simple and lightweight, but must be in flat plane; shims may be needed for this.

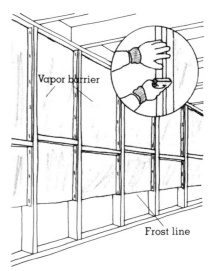

Insulation need extend down only 2 feet below grade, or just below frost line if that is deeper.

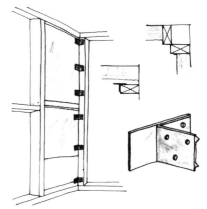

Metal clips offer bearing surface for paneling, allow insulation to be pushed tightly into corner.

HEATERS AND COOLERS

Once your home is weathertight and well insulated, keeping it at a comfortable temperature depends on its heating and cooling system, or systems.

Most American homes have central heating in some form—more often than not a forced air furnace, but sometimes a boiler, heat pump, or some type of radiant heat. Central heat often is supplemented by electric or gas room heaters, a fireplace, or wood stove.

Cooling may come from a central air conditioner, room air conditioner(s), or heat pump.

The owner of any forced air furnace may save by sealing leaky ducts or by insulating ducts that need it. The owner of a gas-fired furnace may profit by having a professional replace the conventional pilot light with an electronic igniter. Relocating or sheltering an air conditioner from afternoon sun can pay real dividends.

However, for most of us, improved energy efficiency in the heating and cooling departments means careful maintenance and watchful use of any existing system, and thoughtful replacement of outworn appliances.

We cite design factors that mean efficiency for all types of heaters and coolers, to help you measure your existing units and to help you shop wisely for new ones.

Savings with Heaters and Coolers

Percentages in the chart assume that the house is weathertight and adequately insulated including at windows.

Action	Range of savings	Your potential savings
Lower daytime thermostat setting from 72°F (22°C) during heating season	Best estimate is 5% of central heater's annual energy use for each degree F, to 65°F (9% for each degree C, from 18°C).	
Raise daytime thermostat setting from 70°F (21°C) during cooling season	Air conditioner use can be diminished about 5% for each degree F to 80°F (9% for each degree C, to 27°C).	
Install setback thermostat (See page 67)	Estimates vary from 5 to 30%; figure depends on how carefully manual thermostat has been used. Ideal setting is 65°F (18°C) days, 60°F (15.5°C) nights.	
Tape leaking joints in warm air ducts (See page 70)	Estimates vary widely. Highest quote is 30% of furnace energy use; most frequent figure is 10%.	
Insulate uninsulated ducting to R-4 (See page 70)	Consensus figure is 10% saving.	
Replace duct air filter monthly (See page 70)	No agreement. Experts figure savings at as little as 2.5%, as much as 10%.	
With gas furnace, replace pilot light with electronic igniter (See page 71)	Estimates range from 5 to 10% savings; consensus trends toward higher figure.	
With oil burner, replace conventional burner with flame retention burner (See page 71)	General agreement that savings will be 5 to 7% of annual oil use.	
With air conditioner or heat pump, provide afternoon shade (See pages 80–82)	Can cut machine's running time about 5%.	
With 17' by 35' swimming pool, add solar or insulating blanket (See pages 84–85)	Can reduce pool heater running time at least 35%, perhaps as much as 50%.	
Add solar heater to pool (See page 85)	Can cut pool heater use by 60%, according to most estimates. Pessimists figure 40% if system is well designed.	

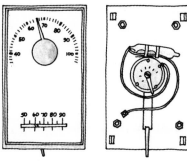

Standard mercury switch thermostat

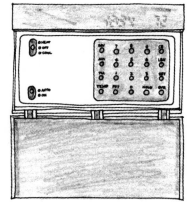

Micro-circuit thermostat

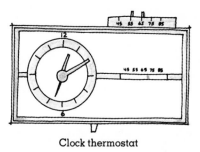

Clock thermostat

Thermostat controls range from manual (left) to automatic (right) to computerized (center).

Thermostats

A thermostat is simply a switch activated by a temperature-sensing device. The thermostat in turn activates or deactivates the heater or air conditioner. It also is the key to efficiency for any central heating system.

Design factors. The most efficient type is the relatively new clock thermostat, sometimes called a setback thermostat.

Clock thermostats allow two or more automatic settings—higher ones for active hours, lower ones for night or periods of prolonged absence. (A comfort bonus comes on chilly mornings when the thermostat turns on the heat before you get up.)

Mechanical reliability has been a problem with some pioneer models, but this aspect of the devices is steadily improving. For people truly forgetful about lowering the settings on manual thermostats, energy savings can range as high as 30 percent.

Most manually set thermostats have an off switch which does away with the need to shut off the furnace itself during prolonged absences. A few elderly models do not have this feature.

To work properly, a thermostat should be about 5 feet off the floor on an interior wall, well away from outside doors, windows, and artificial heat sources. Ideally it should be in the most used room in the house.

All thermostats must be mounted absolutely level to function as designed.

Maintenance. If room temperature varies widely between furnace runs, the furnace is working inefficiently. Look first to the thermostat for the source of trouble.

The principal working parts of a thermostat are the heat sensor, switch, and heat anticipator.

The sensor, a bimetal coil or strip, contracts as it cools, tripping the switch to on, then expands as it warms, tripping the switch to off again. The anticipator heats the bimetal coil so it trips the switch to off just before room temperature reaches the desired level. Dust inhibits the accuracy of both the anticipator and the sensor; efficiency improves with cleaning.

In thermostats with a mercury capsule, clean the entire inside by blowing, or by brushing with an artist's soft brush.

In units with contact points, rub a business card between the closed contacts; then blow or brush dust out of the rest of the mechanism.

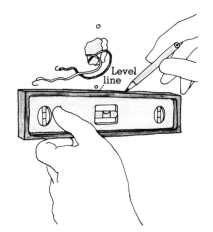

Level line

In replacing a thermostat, be sure wires attach correctly and body is absolutely level.

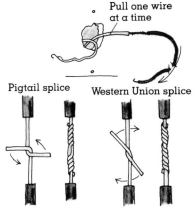

Pull one wire at a time

Pigtail splice   Western Union splice

If wiring must be replaced, use old wires to pull new ones into place. Work one wire at a time.

Zone heating

The single thermostat has one great weakness. It can only sense the air around it, and can only give the furnace one command at a time—on or off.

However, there are ways for people at home to borrow some of the tricks of zone heating—a phrase coined to describe techniques of sending heat to selected areas in large buildings by computer-controlled opening and closing of dampers and vents.

First, as noted in the preceding section, the thermostat should be in the most used room.

Second, rooms not in use should have their heat shut off. This means closing dampers or registers in ducted warm-air systems, or shutting off radiators in steam heated homes. To get the most savings, use sunny rooms more often than shaded ones (vice versa during the cooling season).

Approach the closing of dampers and/or registers of forced air systems with some caution. Too many closed ducts result in back pressure that can cause a furnace to run inefficiently. Check with the manufacturer or a service representative before making major adjustments, especially of damper settings.

A third trick is to seal off unheated rooms from heated ones. In standard buildings this means closing connecting doors (which may or may not require weatherstripping to seal properly). In open plan buildings, closing off unused space may require some structural changes. Sample solutions to typical problems adjoin.

In considering zone heating, look at using individual room heaters as an alternative to turning on a furnace. Though inherently less efficient, room heaters often can heat one or two rooms at smaller expense than a big heater working at a little task (see "Room heaters," page 78).

Shuttering pass-throughs

One of the common architectural devices of our time, the pass-through, allows the passage not only of plates from kitchen to dining room, but also heat. This is fine in the chill of winter, but unkind to guests on warm summer evenings when kitchen heat is an unwelcome extra.

A simple solution is a set of shutters, preserving the traffic pattern while allowing a choice about the heat.

Because space almost certainly will be cramped, a multifold set of narrow shutters is the effective choice, rather than a few wide panels.

Remember that hinge pins should be on the inside of the fold; they alternate sides on multifold installations. Plan from the anchor side outward.

Mount multifold shutters on a top track to avoid sagging (and to keep the pass-through counter clear). The guide pin should be located in the second panel from the center, to allow a small opening while the shutter is in a flat plane, so counter space remains usable.

See the section on shutters (pages 24–28) for insulation values. For shutters to be effective heat shields, the fit must be tight.

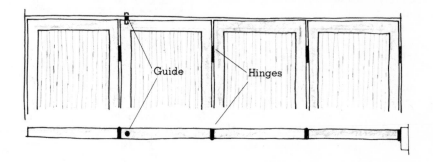

Guide   Hinges

Balcony baffles

Owners of open-plan homes with upper level balconies have unique heating needs, and pivoting panels, or balcony baffles, give them that most energy efficient of all solutions, the flexible choice. With the baffles closed, warm air can be allowed to rise, leaving the lower floor cool and preserving a view and the welcome airiness of an open-plan house.

The drawing here is adapted from two designs. A lightweight set of panels for a 4 by 8-foot opening uses 1 by 4-inch framing, an insulating core of polyurethane foam, and gaily painted metal covers. A heavier set uses 2 by 4-inch framing, an insulating core of fiberglass batts, and wood paneling to match adjacent walls. The opening is 7 by 10 feet.

The key to tight fit is the parallelogram, as shown. Rectangular edges cannot meet precisely, or they collide when rotated.

The vertical pivot for each panel should be off-centered so that the panels swing outward and do not intrude on living space. One of our models had the pivots at one edge so the whole panel swung into space above the lower level. The other model was offset so that two-thirds of the panel hung in space, while slightly less than a third stuck into the adjoining bedroom.

Of course, using more numerous, and therefore narrower, panels decreases the overhang problem.

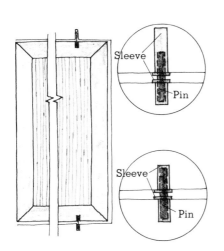

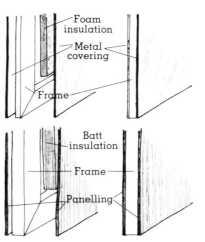

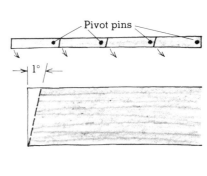

Warm air furnaces

Perhaps the most common residential power plant is the furnace, which heats air and then distributes it throughout the house via a system of ducting. Most of these heat with a flame fueled by natural gas, propane, oil, or, in a few cases, wood or sawdust. In some parts of the country, electric furnaces heat air with resistance coils, but several electricity-poor states discourage their installation.

The prime measure of furnace efficiency is the heating capacity as it relates to house size. Sometimes called heat output or bonnet capacities, furnace heating capacities range from 21,000 to 180,000 BTUs per hour. The rule of thumb for measuring minimum heating capacity follows:

First, calculate the number of square feet of heated floor space (do not include the garage or other unheated space). Then multiply this figure by 40 if the lowest expected outdoor temperature in January is 30°F, by 50 if the lowest predicted temperature is 20°F, and so on, adding 10 to your multiplier for every 10°F decrease in the expected low.

If your home is fully insulated (ceiling, below the floor, and walls), subtract 25 percent from the result, or add 25 percent if your home is insulated about as well as the average barn.

The final figure should equal roughly your furnace's heating capacity measured in BTUs.

To operate at peak efficiency, all furnaces—correctly sized or not—should have a reliable clock thermostat and tight, well-insulated ducting. Most should have flue dampers. Other efficiency factors, particular to each type of furnace, are noted in individual descriptions.

First we discuss ductwork

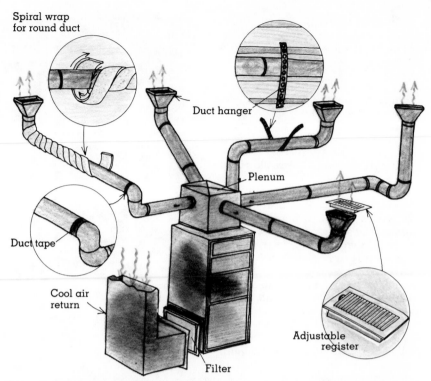

Tape every joint in ducting, then tape every seam in insulation applied to rectangular ducting; with spiraled insulation, tape ends only.

and flues, common to all warm air furnaces. Discussion of individual furnace types follows.

Ducts and filters

Open-flame and electric furnaces send heated air to the house and bring cooled air back for heating through an all-too-fallible network of ducts. Owners of these ducted systems can do much more than keep the filter in the cool air return clean.

Filters. Temporary types should be replaced and permanent ones washed clean as often as they show clogging thicknesses of dust. For most of us, this is about once a month. Filters in most models are located where the cool air return reaches the body of the furnace.

Most filters are designed for air flow through them in the direction of arrows printed on the frames; they must be installed with the arrows pointing in the direction of the air flow.

Ductwork. If you have a poorly assembled system, you can save as much as 40 percent in energy usage by making it airtight (provided your filters are clean).

Tape joints and cover with insulation. For rectangular ducts, use batts lengthwise, taping all seams with duct tape. For round ducts, use spiral insulation as shown above. If your ducts carry cold, conditioned air, use insulation with a vapor barrier, barrier side out, to prevent condensation on the duct.

Dampers. If you find yourself constantly adjusting registers to make rooms more comfortable, rebalancing dampers may be profitable. The job is best done by a professional—poorly balanced systems can develop back pressures that can burn out the firebox in your furnace.

If your system has no electrically driven dampers, adding them could repay the cost of the job—that is, if the house structure provides no major obstacles. Again, the work is for a professional.

Blowers

Except for old-fashioned gravity-flow furnaces with air movement that is inefficiently sluggish, all furnaces incorporate a blower to move air past the heat exchanger and through the ducts.

The blower, sometimes called the squirrel cage, is located in the return-air plenum near the heat exchanger. The assembly consists of a motor, a fan, and, usually, a drive belt (some fans are attached directly to the motor). Good performance depends on regular maintenance of all three.

Focus first on the fan. Dirty blades will not move the fan's rated volume of air. You may need to unscrew and slide the unit all the way out of the blower compartment before cleaning. Usually brushing is enough, but an uncommonly dirty fan may need washing in detergent. Many fans have balancing weights clipped to them; take care not to move the weights while cleaning.

Also check pulley alignment and belt tension regularly. Use a straightedge to check pulley alignment. If it's out, loosen the locknuts on the pulley's shafts and adjust it. Be sure to retighten locknuts securely.

The belt should deflect 1 inch with modest thumb pressure midway between the pulleys. If it is too loose or tight, adjust it by loosening the set screw on the motor housing.

If the belt is worn, or has hardened and cracked, it should be replaced. To minimize wear, spray on belt dressing at the beginning of each heating season.

Most blower motors have oil cups, but some are sealed. If yours is sealed, lubricate it by squirting five drops of lightweight oil on the felt pad that cushions the shaft where it enters the motor. Oil other motors by squirting five to ten drops of lightweight oil into each cup.

Follow lubrication instructions posted on the furnace.

Flues

Also called an exhaust vent or stack, the flue carries the products of combustion—water, CO_2, soot, creosote, and poisonous gases—away from an open flame furnace and discharges them into the outside air, either directly or via an existing masonry flue.

Flues wear out through corrosion. To check for this, tap the pipe in several places while the flue is cool. If it sounds dull rather than giving a metallic ringing, press the pipe inward. If it doesn't spring back, it needs to be replaced.

Gas-fired furnace

Whether fueled by natural gas or propane, this furnace has a burner unit, a heat exchanger, and a blower. The burner unit has rows of pinhole ports aimed at the heat exchanger. The flaming burner heats the exchanger, which in turn heats air blown past it by the blower.

All three parts must be kept clean for the furnace to operate at peak efficiency.

Design factors. Little can be done to affect performance of the appliance itself, save for replacing a conventional pilot light with an electronic igniter.

Some engineers suggest that an outsized burner can be replaced with a smaller one to save fuel. This option, called down-sizing in the trade, is rejected by some experts on grounds that the other elements in the system are thrown out of balance, resulting in a net loss of efficiency.

Maintenance. Keep surfaces around the pilot and burner assembly free of dust and soot. Use a pin or straightened paperclip to dislodge soot in burner ports. At least once a year, wipe or vacuum the heat exchanger surface, because grime reduces its ability to conduct heat to the air.

The time for all of these tasks is just before you relight the furnace for the new heating season. At any other time, be sure the furnace is off before beginning.

Keeping the pilot properly adjusted saves energy. A healthy pilot burns steadily and has a blue flame with a hint of yellow at the tip. If the flame is weak and mostly yellow, or if it is a jumpy blue and noisy, adjustment is required. Some manufacturers provide adjustment instructions, which usually involve turning a set screw. If in doubt, call for professional service.

Oil-fired furnace

More complex than its gas-fired cousin, this furnace's burner mixes oil and air. A spark ignites the mixture, and a small blower within the burner whips the flame into the furnace combustion chamber. (The motor that powers the blower also powers a fuel pump.)

Design factors. The most efficient oil-fired furnaces have a device called the flame retention burner. Contact your furnace manufacturer to find out what kind of burner you have. You may be able to replace it with a flame retention burner, if you don't already have one. The work should be done by a professional. Fuel use can be reduced as much as 30 percent.

Another option to reduce fuel use is derating—the replacement of the oil/air nozzle with a smaller one. Also a job for a professional, derating can re-

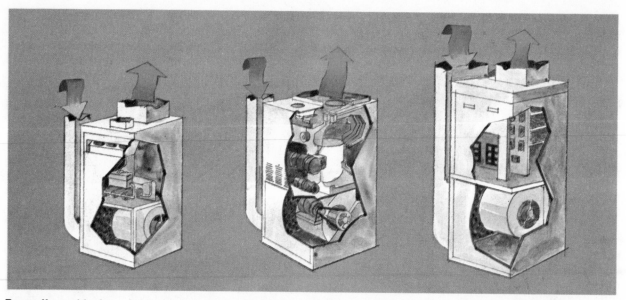

Regardless of fuel used, forced air furnaces function much alike. A fan pulls cool air through a filter into the furnace body, then blows it past the heat exchanger. From there, ducting takes warmed air back into the house. From left to right, furnaces are gas-fired, oil, and electric.

duce fuel consumption as much as 15 percent.

Before committing to either modification (or to buying a new furnace), have a service technician perform the smoke and CO_2 tests to determine the efficiency of an existing burner.

Maintenance. As with gas-fired furnaces, vacuum or wipe clean the heat exchanger (located at the top of the combustion chamber). Also clean blower fan blades.

To check for a tight seal of the mounting plate between the combustion chamber and blower, pass a cigarette lighter around the rim of the plate while the burner is on. If the flame is sucked toward the plate, tighten the bolts.

Keep the oil strainer clean. The symptom of a dirty one is a sputtering flame as the burner turns off. Before cleaning, turn off the main switch and then close the valve in the oil line. Disengage the strainer cover plate and remove the strainer screen. Clean it by rinsing in kerosene. Replace, making sure to tighten the cover plate firmly.

Lubricate the motor shaft

with two drops of oil at the start of each heating season.

Electric furnace

This furnace heats with two to six resistance coils. Though expensive to operate in all but a few electricity-rich regions, it needs virtually no maintenance.

Boilers

The refrain "I got steam heat" applies to fewer of us today than to past generations. Yet, particularly in the Midwest or East, steam and hot water systems continue to keep us warm.

Design factors. You can buy new boilers that are fired by oil, gas, or electricity. Their heating capacities range from 55,000 to 350,000 BTUs (British thermal units) per hour. Use the formula on page 70, developed for furnaces, to determine how many BTUs your own boiler should put out. Then look for design factors that save energy.

A "wet-base boiler" is partic-

ularly efficient because water encompasses the combustion chamber. Another energy-efficient type, the "multiple pass boiler," recirculates combustion gases through the heat exchanger. A third design feature that boosts efficiency is the fuel-frugal flame retention burner mentioned on page 71.

Maintenance. Boilers work pretty much as furnaces do (see page 70), except that watertight copper or galvanized pipe replaces air ducts. In forced-water boilers, a circulating pump replaces the air blower. Traditional cast-iron radiators, or the more efficient convectors and baseboard radiators (which employ a series of fins surrounding exposed pipe sections) heat by carrying hot water or steam into the rooms of your home.

Tune-up tips pertaining to a furnace's components—to the thermostat (page 67), the burner (page 71), and the flue (page 71)—pertain to boilers, too. You'll also want to oil the circulating pump's shaft, and tighten the shaft's packing nut to stop leaks. If the relief valve is dripping, replace it.

A boiler's pipes and radiators (or convectors) deserve special care. Wrap pipes that pass through unheated areas, such as the attic or garage, with unfaced R-11 insulating batts or blankets (see page 49 for meaning of R-value).

Paint radiators with paint formulated for radiators (not the standard metallic paint); where you can, fasten a sheet of aluminum foil behind radiators to reflect heat into the room. Vacuum dust from exposed metal surfaces of both radiators and convectors.

To increase a boiler system's efficiency, a service technician can retrofit your flue with an electrically controlled damper and a secondary heat exchanger.

Each year, before the heating season begins, have a service technician drain and clean the boiler. Then, if yours is a hot water (not a steam) system, bleed air from your radiators. Open each air release valve, usually located at the top of the radiator, while holding a bucket under it. Leave the valve open only until water comes through.

Elements of an efficient fireplace: Ducting for convection-warmed air; outside combustion air; airtight glass door; ability to radiate heat to fireside sitters.

Wood Heat

In this era of ever higher utility rates, a growing number of homeowners are turning to wood for supplementary or even alternative heat, wherever supplies are plentiful enough for the price to be competitive.

Wood stoves offer the more practical source of heat, but fireplaces have an inextinguishable spark of romance about them. Both are gaining in popularity. And in both cases, more savings come from wise original purchase than from modification of an inefficient model.

Usable heat from a fireplace ranges between 10 and 30 percent of all the heat its fire creates; wood stoves give about 40 to 60 percent. In absolutes, the range for both is 10,000 to 120,000 BTUs per hour. This compares with 100,000 BTUs from a therm of gas or from 30 kwh of electricity.

If you're thinking seriously about buying one or the other, consult one of these *Sunset* books: *Homeowner's Guide to Wood Stoves* or *How to Plan and Build Fireplaces*.

If you already have a stove or fireplace, the following may help you gain a bit more heat from the wood you burn.

Fireplaces

Odd as it may seem, a roaring fire in a poorly designed fireplace can leave a drafty house colder than no fire at all. As the fire draws oxygen to feed itself, it can pull in more chill from outside air than it radiates heat, leading to a net drop in indoor temperatures. A standard fireplace produces as usable heat about 10 percent of all the heat its fire creates; an efficient one may reach 30 percent. The rest of the heat is lost mostly out the flue.

Design factors. An efficient fireplace design, like that shown on page 73, uses outside air for combustion, sends air warmed by convection into the room (perhaps with a boost from small fans), minimizes drafts with glass doors, and radiates heat to more than one side. In short, an efficient fireplace is the closest possible imitation of a wood-burning stove.

If your fireplace lacks these characteristics, you can add two of them. Commercially available convection grates give out a good volume of warmed air, especially when a fan speeds the airflow. Tempered glass doors cut down on drafts and slow the combustion rate of the fire, but they also trim a slice off the amount of radiant heat given off by the fire.

The radical solution is to abandon the fireplace in favor of a wood stove connected to the fireplace flue. The sketches below show how this is done.

Maintenance. The principal job is keeping the flue clean (see drawing at right).

Operating tips. Burn seasoned wood only.

Close the damper tightly when no fire is burning. (Warm air escapes up the flue whenever the damper is open.)

If your fireplace draws poorly, even with a clean flue, have a fireplace contractor elect to do one of the following: decrease the volume of the firebox by adding a layer of firebrick to its floor; add to the flue a rotating wind cap that turns to keep wind from entering; or extend the flue.

If your fireplace worked well until new caulking and weatherstripping tightened the house, you may need to open the nearest window a crack to restore lost oxygen supplies.

Wood stoves

Because the thick metal sides of wood stoves heat the air by convection, they are much more efficient than fireplaces, which only warm objects by radiation. (See Glossary, page 94, for definitions of convection and radi-

Cleaning a flue

Cleaning a flue is simple, but messy, as shown in drawing. One warning: Do not use household vacuum to fine-clean firebox; motor will clog. Use shop vacuum, or go slowly with brush and broom.

Replacing a fireplace with a stove

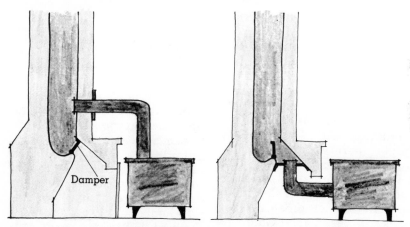

Damper

Sealing off an old fireplace with a metal panel allows its forehearth to hold a wood stove, provided the forehearth is large enough (or is enlarged) to meet code requirements. In replacing a fireplace with a wood stove, it is absolutely imperative to carry stovepipe as far as fireplace damper, or farther. Venting stove directly into old firebox can create an actual bomb, because creosote will build rapidly in closed space. Once ignited, creosote is likely to explode.

ation.) The efficiency range is a usable 40 to 60 percent of the total heat a stove's fuel produces.

Size and other design factors give stoves an enormous range of absolute heating power. The spread is from 10,000 to somewhat more than 100,000 BTUs per hour. (The figures are inexact because not all wood gives off the same amount of heat, and no fire burns at a constant intensity for hours on end.)

Design factors. Good performance in a wood stove means getting the most complete combustion possible. Two features that contribute to this are airtightness and baffles; the former prevents the escape of air and the latter retards it. They thereby allow temperatures to build, leading to more complete combustion.

Maintenance. Keep door gaskets and joints sealed. Gaskets must be replaced to prevent leaks. Leaks in the joints of cast-iron stoves occur as old cement hardens, cracks, and falls out. Apply stove cement to leaks (follow instructions accompanying the cement).

Keep ashes at a level that does not restrict airflow in the firebox.

To minimize rust, keep stoves well painted with stove black or high-temperature paint to match the original finish.

The wood stove flue is cleaned exactly as a fireplace flue, except that the stovepipe connecting it to the firebox must be disassembled and taken outdoors for cleaning.

Operating tips. When no fire is burning, keep dampers tightly closed to prohibit warm air from escaping up the flue.

Burn only seasoned wood. Choose the species that give the most heat for the dollar. (See the chart for heating capacities of several kinds of wood.)

Comparative heat values of woods

Woods vary widely in the heat they produce when burned. These figures assume the wood has been seasoned to 20 percent moisture content (about nine months drying time), and that the cord is a full 128 cubic feet (4 feet by 4 feet by 8 feet).

As comparison, remember that one therm of gas (100,000 BTUs) gives 80 percent usable heat in an efficient furnace, and that 30 kwh produces 90,000 BTUs.

Fuelwood type	Available heat in millions of BTUs at efficiencies of			
	10%	30%	40%	60%
Apple Hickory, shagbark Ironwood (hardhack) Black locust Live Oak	2.4	7.2	9.6	14.4
Dogwood Rock elm White oak	2.3	6.9	9.2	13.8
Yellow birch Sugar maple Red oak	2.1	6.3	8.4	12.6
White ash Eucalyptus (all species) Black walnut	2.0	6.0	8.0	12.0
White birch Black cherry Douglas fir Red maple Tamarack	1.9	5.7	7.6	11.4
Green ash Pitch pine American sycamore	1.8	5.4	7.2	10.8
Black ash American elm Silver maple	1.7	5.1	6.8	10.2
Alder Cedar Hemlock Red spruce	1.4	4.2	5.6	8.4
Aspen (poplar) Red pine	1.3	3.9	5.2	8.4
Basswood Balsam fir Ponderosa pine Sugar pine White pine Redwood	1.2	3.6	4.8	7.2

THE POSSIBILITIES OF ACTIVE SOLAR HEATING

Typical active systems employ collectors to heat air or liquid, which is forced to the storage mass. Heat is then fed into distribution system of conventional central heater. When temperature in storage mass dips too low to help warm house, dampers or valves close off solar side of system, and conventional heater works alone, as in illustration above right. Collectors may be on roof, on wall, even on ground.

Active solar space heating is a tempting alternative to conventional furnaces for people living in houses that don't readily lend themselves to passive solar heating.

Though ingenious advances in design make active solar heating ever more possible, there are still some big ifs.

The primary one is cost effectiveness. The second is difficulty of installation. Active solar systems are in a curious position in this latter regard. They can be installed in many existing houses without major remodeling. However, the complications are such that professional installers must do the work in nearly every case.

Though active solar space heaters fall outside the scope of this book of do-it-yourself projects, we nonetheless offer the following brief survey to help you think about this alternative energy-saver. For a more detailed explanation of the applicabilities of solar space heating, see the *Sunset* book *Homeowner's Guide to Solar Heating & Cooling*.

The basics

The object of any solar heating system is to gather heat from the sun and store it until needed.

Active solar installations consist of collectors, ducts or pipes to transport heated air or liquid, and a storage area. The active (moving) components are fans or pumps, dampers or valves, and thermostats.

Radiant solar energy becomes heat when it strikes the blackened surface of a collector, which then raises the temperature of the air or liquid it contains. Even on a freezing winter day, direct or moderately filtered sunlight can raise the temperature in such a solar collector to over 100°F/ 38°C.

Air and liquid are about equally effective mediums to collect, transport, and store heat.

Air systems. An air system requires relatively simple hardware and won't leak, freeze, or corrode. But it has one major disadvantage: its huge space requirement. The ducts, larger than an average wall thickness, are difficult to install in an existing building; the usual storage mass in whole-house systems is many tons of rock. (Not uncommon is a 50-ton mass requiring a bin 10 by 30 by 4 feet.)

As a result, an air system is almost impossible to add to an existing house in the familiar design with

Solving many problems of adding active solar heat to existing house, wall-mounted collector contains large ducts that would pass through building in designs using roof-mounted collectors. Baffles slow rise of heated air; central column returns hot air to storage mass directly below. Regular furnace ducts carry heat from storage mass to rooms. (Vents in wall also permit use of collector as a passive heater; see pages 44–45.) Drawing is from J. K. Ramstetter design called "Winter Star."

the collectors on the roof. But where the storage mass can be put in a basement and the collectors placed on the ground or on a wall, as in the example shown above, air is a possibility.

As storage masses become more space-efficient, air systems will become more readily installable. Substituting an efficient heat-holding material called "eutectic salts" for rock is one promising experiment toward this goal. The salts are contained in enclosed trays that stack.

Liquid systems. These require complex plumbing and hardware but less than half the storage volume of an equivalent air system.

The pipes can be fitted into walls without disturbing the appearance of finished rooms. Another advantage is the slightly greater efficiency of liquid systems in heat collection and dispersion.

However, the dispersion advantage may lie with your existing heating system. If you already have a forced air furnace, ducting for dispersion is in place. Similarly, if you have a boiler and radiators, the solar heat can be distributed through that system.

Passive solar space heating and solar water heating are more attainable goals for do-it-yourselfers. For examples of the former, see pages 44–45; for a primer on solar hot water, see pages 88–90.

Room heaters

In principle, room heaters are less efficient than central heaters, because they put too much heat into too small a space. However, when you want to warm one cozy nook, using a room heater instead of a central heater saves energy.

Room heaters come in hundreds of designs but only three main types.

Electric resistance. The group includes electric baseboard radiators, and portable or wall-mounted coil heaters. The radiators heat with a resistance coil hidden inside a finned heat exchanger; the coil models have an exposed coil that heats directly.

Efficient electric models have thermostats and fans. Coil types should have large chrome reflector surfaces; radiators should have a large heat exchanger surface. Portable units use 110 volts, while built-in models use either 110 or 220 volts. The latter is preferable where there is a choice. The rule of thumb in choosing a convection heater is 1 watt per cubic foot of air volume in the room.

All electric resistance heaters warm room air by convection, so are often called convection heaters.

Quartz heaters are a subtribe of electric heaters. They have a coil, which heats a quartz tube by convection; the tube then heats objects quickly by radiation (see Glossary, page 94). They are most effective at 3 to 10 feet, and are best used in large, hard-to-heat rooms where the user can huddle close.

Kerosene. These portable heaters have either an enclosed or an exposed flame. The former heats space by convection. The latter heats by a combination of

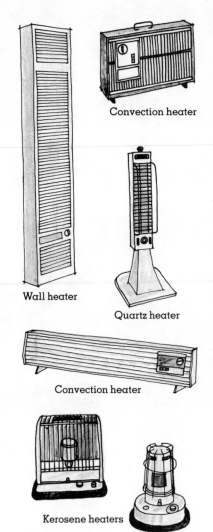

Convection heater

Wall heater

Quartz heater

Convection heater

Kerosene heaters

convection and radiation—the large exposed flame heats objects first. BTU output ranges from 8,000 to 20,000 per hour. No venting is required.

Wall heaters. In effect, these are miniature gas or oil furnaces. Designed to fit between wall studs, they pull cool air in through a duct near floor level and push hot air out through a second duct near the top of the unit. There is no other ducting, only a flue to expel gases.

Wall heaters are most often installed in single-room cabins, or in rooms added during a remodel.

For hints on maintenance and operation, see "Warm air furnaces," starting on page 70.

Heat pumps

As a home heater, the heat pump seems to achieve the impossible: it moves heat from the cold outdoors to the warm indoors. In the process it seems to produce more energy than it consumes.

To achieve this apparent miracle, the device just picks up and moves sun-generated heat energy already present even in cold winter air. This can be far less expensive than creating heat with electricity. Under ideal conditions, the coefficient of performance (the ratio of heat output to energy used) of a heat pump can be as high as three to one. In comparison, a perfectly efficient electric heater's coefficient of performance is one to one.

The heat pump is actually a central air conditioner that can turn itself around to heat. The system is easier to understand if you remember three principles of physics:

• Heat energy flows from a warm place to a cooler place.

• When caused to expand, gases become colder.

• When compressed into a tighter space, gases become hotter.

Like an air conditioner (and the kitchen refrigerator), the heat pump employs a closed loop of refrigerant gas. The difference is that the gas circulates through not one but two coils—one inside the house, the other outdoors. (Though they're called coils, the heat exchanging devices are finned more like a car radiator.) Each coil has a blower fan to move air across its surface.

The diagrams show how the refrigerant flows in one direction to heat, and in the reverse direction to cool.

Attractive as the basic idea seems, payback is questionable in any area where air conditioners are not normally re-

How heat pumps heat . . .   . . . and cool

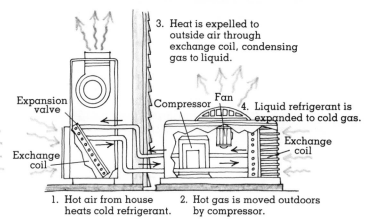

4. Heat is transferred to air through exchange coil in duct. Heat loss causes vapor to condense to liquid.

1. Liquid refrigerant is expanded to cold gas.

Expansion valve

Exchange coil

2. Heat from outside air moves into cold gas in exchange coil, boiling gas into warm vapor.

Fan

Compressor

Exchange coil

3. As vapor is compressed, it becomes hotter and is pushed indoors.

3. Heat is expelled to outside air through exchange coil, condensing gas to liquid.

Expansion valve

Exchange coil

Compressor

Fan

4. Liquid refrigerant is expanded to cold gas.

Exchange coil

1. Hot air from house heats cold refrigerant.

2. Hot gas is moved outdoors by compressor.

quired, because the built-in air conditioning capacity of the heat pump makes it far more expensive to buy than a furnace alone.

Also, the heat pump is no more efficient in cooling than any standard air conditioner. Potential buyers should therefore look only at potential heating savings, which can be significant in the right climate.

The other climate consideration is winter cold. A heat pump may not deliver adequate heat when outdoor temperatures dip below 35°F/2°C, and backup electric resistance heating may be required in the ductwork. (One alternative where relatively inexpensive natural gas is available is a hybrid heat pump system grafted onto an existing gas furnace.)

Design factors. There are two basic types of heat pumps: air-to-air and water-to-air. The latter can be tied into swimming pool solar heaters to excellent effect, but air-to-air heaters are the most versatile and practical type.

Like air conditioner size, heat pump size is measured in terms of tons of heat removed from the house in an hour. A ton of heat is equal to 12,000 BTUs (see Glossary, page 94, for BTU).

Where winter temperatures stay above 35°F/2°C, a 3-ton unit often is enough for a 1,500-square-foot house.

The following features aid heat pump efficiency:

• A *filter drier* is a stick of pumice in a metal housing with a little tube poking out each end. It relieves the refrigerant of moisture that can form damaging hydrochloric acid.

• An *accumulator* stores excess refrigerant, reducing strain on the compressor.

• A *crankcase heater* helps keep compressor valves warm enough to function properly.

• An *outdoor-coil temperature sensor* reverses the heat pump flow to defrost the heat exchanger when required.

• An *outdoor thermostat* helps keep the backup heater dormant until outside temperatures drop below a set point, usually 35°F/2°C.

The heat pump often has larger ductwork and more registers than other central heaters. This is because the heat delivered reaches the room at about 100°F/38°C, compared with 125°F/52°C for furnaces. If this relatively cool air is forced through small ducts, it speeds up and feels slightly chilly, like a draft. It is therefore a good idea to equip existing registers with deflectors, which steer air

away from occupants and minimize this effect.

Operating tips. Effective insulation is important because of the low temperatures of heated air.

Install the outside unit away from bedrooms or other rooms where its blower noise will be a bother. (Remember the neighbors in this connection.)

Protect the unit from standing water, snowdrifts, and strong winds, but allow for free air circulation around it.

To get the most from sun heat, place the unit on the south or west side of the house. (This can diminish its effectiveness as an air conditioner unless removable sun shades are provided. See pages 81–82 for suitable sun shades.)

Maintenance. Once a month, change the filter in the return air duct near the indoor heat exchanger. (See manufacturer's instructions.)

Clean the outdoor heat exchanger once a year in late spring. After turning the main power switch off, unscrew and remove the top grille. Hose off the heat exchanger fins. Clean all dirt from the base pan. (Keep water off fan motor.)

At all times, keep the unit free of fallen leaves and other debris.

Air conditioners

The advent of the mechanical air conditioner during the 1950s turned summer from a long misery into a productive season in the humid southeastern United States, and made it much more comfortable in most of the rest of the country.

Valuable as the machines are, they also use a major portion of home energy budgets wherever they do their good work.

Air conditioners cool and dehumidify air by drawing it past a heat exchanger (finned tubing containing cold refrigerant—kept cold by circulating through a compressor-condenser loop much like those in refrigerators.)

The cooled, dried air is blown back into the house, having been cleaned by a filter along the way. The heat is exhausted to the outdoors.

There are two basic types, the room air conditioner and the central air conditioner. The former cools only the area in which it is located. The latter cools the whole house through a system of ducts (usually the same ones used by the central heater).

Air conditioners are rated according to their capacity to remove heat, measured in tons per hour. A ton of heat is equal to 12,000 BTUs (British Thermal Units).

For both room and central air conditioners, the rule of thumb is one ton of heat removal capacity for every 350 to 500

Window-unit shade

This simple shade for window-mounted room air conditioners is attached either to window trim or through the exterior wall surface to studs alongside the window.

It also is meant to be removable in winter, whether the air conditioner is taken out of the window, or left in place and covered with a shield to prevent air infiltration through the grilles.

To keep the shade lightweight, it is made entirely of 2 by 2-inch rough redwood or other softwood. It also requires two 90° angle clips, and slot-head or lag screws to attach the angle supports to the wall.

In buying wood members, plan for at least 1 inch clearance between the top of the air conditioner and the bottom of the shade. The shade should extend 1 foot past the outer end of the unit; the baffles can extend as far to either side as you choose or as sun angles dictate.

The baffles should be spaced ½ inch apart for maximum shading and adequate top ventilation.

Here is the order of assembly: Cut supports and knee braces to length. Drill screw holes in supports for angle clips; drill holes where knee braces meet the wall and the supports (see diagram)—do not secure any members yet.

Nail baffles to supports, then secure braces and angle clips to supports with screws.

Finally, have a helper hold the shade in position while you secure clips and braces to wall, starting with clips.

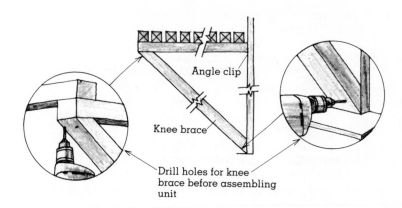

Angle clip

Knee brace

Drill holes for knee brace before assembling unit

square feet of floor space. Formulas can become much more precise, taking into account climate and other factors.

When possible, air conditioner units should be placed on a north wall. If a unit cannot be placed on a shady wall for whatever reason, it should have a sun shade. Sun shades reduce energy use as much as 5 percent by shortening the air conditioner's running time. A sun-heated unit must cool itself before it can begin to lower the temperature of hot indoor air.

Room air conditioners

These units range in cooling capacity from ⅓ ton to 2½ tons (4,000 to 30,000 BTUs) per hour. They are most useful if you wish to cool only one or two rooms, if your central heating has no ducting system, or if you have added a room not served by existing ductwork.

Design factors. Thus far room air conditioners are the only appliances carrying federal Energy Efficiency Ratings (EERs). These ratings show by means of a single number the ratio of energy input to productive output for devices of a type. The higher the number, the more efficient the device.

Room air conditioner EERs range from 5.1 to 10.7. Energy-efficient features that raise EER numbers are large condensers, fans that can run with or without the compressor, and 220-volt (rather than 110-volt) current usage.

Maintenance. At the start of the cooling season, vacuum the

Lean-to shade

Screening the outdoor portion of your heat pump or central air conditioner makes sense in two ways. In addition to keeping the sun at bay, a well-designed screen can dress up a piece of equipment not designed to win any awards from landscape architects.

If you're covering a heat pump, the shade should be removable, because in winter, exposure to sun heat increases the unit's heating efficiency (see "Heat pumps," page 78).

Shade screens must let heat escape freely. There should be 4 feet of clear air space above the unit, 2 feet of space on the access panel side, and at least 1 foot on all other sides.

For this detachable, fold-up shade, you need two 12-inch square concrete pads, two 3-inch lengths of steel reinforcing bar; enough 2 by 2-inch rough redwood or cedar for the frame, the baffle strips, and ledger; and 5/16-inch carriage bolts with wing nuts for all joints.

Using a masonry bit, drill holes in concrete pads and in upright supports for the reinforcing bar. Epoxy the bars into the pads *only.*

Uprights must be at least 4 feet longer than the unit is tall; supports must extend 32 inches beyond the outer edge of the unit.

If the ledger you attach to the house is about as wide as the machinery, cut the baffles so they overhang 24·inches on each side; space them ½ inch apart.

Attach the cross supports to ledger and posts with bolts.

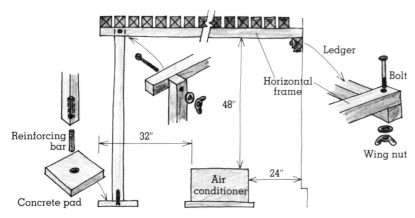

condenser and evaporator coils, and clean the drain tube (see manufacturer's instructions). If replacing a unit in a window, make sure it tilts slightly toward the outside, to keep water from dripping on the floor.

Operating tips. Place the unit in the wall receiving the least direct sunlight. If it must be on a sunny wall, build a shade for it like that shown on page 80. (Sun heat on the unit diminishes the condenser's capacity to dissipate heat from indoors.)

Seal all edges around the unit. If it is permanently mounted in a wall, caulk the joints between the air conditioner body and both exterior and interior wall coverings. If the air conditioner is mounted in a window, be sure its side wings fit snugly, or, if it has none, that ¾-inch plywood or other blocking is in place. Use foam gaskets or conventional weatherstripping to seal all cracks.

For permanently installed units, keep the outside grille covered during the winter to prevent winds blowing through the housing. You can buy commercial covers or make your own.

Central air conditioners

Central air conditioners cool air just as room air conditioners do, except that they push the conditioned air into ducting for distribution throughout the house rather than blowing it straight into a room.

The machinery is on a grander scale. In some models all of the components are housed together in an outdoor unit. In others, the evaporator coils adjoin the forced air furnace in order to send cold air directly through its ducting, while the rest of the machine is outside. Capacities range from 2 to 5 tons (24,000 to 60,000 BTUs) per hour. For a huge house, commercial models offer still larger capacities.

Design factors. Central air conditioners range widely in efficiency. Some effective units accept evaporative precoolers that cool the air on its way into the air conditioner. Two-speed condenser fans and compressors save energy by operating at the costlier high speed only when indoor temperatures exceed a set limit, usually 85°F/29°C.

Maintenance. Replace the disposable filter, or clean the permanent one, monthly. At the start and end of each cooling season, hose the outdoor condenser coils clean *after* turning off the main power. Take care not to get spray into the motor housing. Keep leaves and other debris out of the grille.

Operating tips. Shade the outdoor housing, if located where sunlight strikes it directly. (See adjacent and page 81 for examples of sun shades you can make.)

Keep the ductwork airtight and well insulated, remembering that the vapor barrier on insulation must face outward to avoid condensation. (For more about vapor barriers, see page 51.)

Freestanding shade

When a central air conditioner or heat pump is located some distance from a wall (usually to muffle fan noise), shade must often be made.

This shade is a freestanding variation of the one shown on page 81. It also collapses so a heat pump can be exposed to winter sun heat.

The frame, made entirely of 2 by 2s, rests on pins and pads, as does its attached cousin.

It differs in being covered by lath, which can be laid flat with spacing or can be set at any angle to increase air flow with no loss of shade.

The key to its collapsibility is in the top corners, where joist hangers provide bearing surfaces and interlocks for the sides; they must be drilled and bent as shown.

Controlling humidity

Though many of the effects on the energy bill are indirect, controlling humidity inside the house can pay real dividends.

In summer, comfort comes with humidities ranging as low as possible. When temperatures reach 90°F/32°C, comfort begins at 50 percent relative humidity (RH), but lower feels even better. (The ability of the atmosphere to hold moisture is greatest when air is hot, least when it is cold, so comfort depends on how much moisture the air is holding relative to its capacity rather than any absolute amount.)

Comfortable living in winter is enhanced when the RH ranges around 50 percent. For normal indoor temperatures, dry air feels chillier than moist air, so a room can be as comfortable at 65°F/18°C and 50 percent RH as it can at 70°F/21°C and 40 percent RH. Air moister than 60 percent RH may be comfortable, but it favors the bacteria that cause respiratory problems as well as those that produce molds, mildews, and related pests.

People who live in regions with hot, steamy summers can use dehumidifiers to reduce the burden on air conditioners or other coolers. Residents of regions with mild, moist winters can use the same devices to reduce moisture damage to wood and other materials.

Conversely, those who live where winters are dry and cold benefit most from humidifiers in that they decrease the need for heaters.

Dehumidifying

A family of four generates almost 16 pints of moisture a day, most of it as a by-product of breathing, and the rest from cooking, laundering, bathing, and related activities.

If outdoor humidity and temperatures peak near the top of the comfort range, extra humidity usually can be dispelled by turning on a kitchen or bathroom fan vented to the outdoors.

In regions of relentlessly high temperatures and humidity, a mechanical dehumidifier may be needed and may pay for itself in a short period. Such devices are common in the southeastern United States, especially the Gulf Coast states, but may yield dividends elsewhere.

Dehumidifiers also can play a role in the rainy Pacific Northwest, where weathertight houses can become so moisture-laden that mildew and other decay will develop.

Besides lowering humidity, dehumidifiers generate heat. Proponents say that, at 970 BTUs per pound of moisture condensed, dehumidifiers more than pay their own way on the energy bill in moist-winter areas, provided they can be located so that generated heat will feed into the living space.

The machines channel air past a chiller, which produces condensation. The dried air returns to circulation, and the extracted water drains into a pan.

Design factors. Dehumidifiers draw 300 to 500 watts of power per day, and they remove 12 to 40 pints of water from the air in that time. They are rated on this moisture removal capacity (which should be certified by the Association of Home Appliance Manufacturers). They must be linked to a humidistat, a switching device that measures moisture much as a thermostat measures heat.

Maintenance. If mold grows on the condensing coil, scrub it off with a brush; then wash the coil with detergent. Vacuum lint from grilles and replace the dry-air filter monthly.

Operating tips. Settings usually include Extra Dry, Dry, and Off. Use the Extra Dry setting only in times of distinct discomfort.

Do not operate in room temperatures below 65°F/18°C, or the condenser will ice up and become ineffective until defrosted.

Humidifying

The cheapest, simplest humidifier known is a pan of water balanced on a radiator, register, or similar heat source. The method is perfectly effective, though it requires regular watching to keep the pan from evaporating dry. In areas where there is only an occasional cold, dry spell, a pan or two in the most-used room may be enough.

As noted earlier, the house is a daily source of moisture, which usually is vented outdoors. Electric clothes dryers can have their moisture returned to the house, if they operate in part of the heated space. The vent tube can be disconnected from the housing that guides it through an outside wall (and the housing covered to prevent air exchange), or the tube can be fitted with a commercially available directional valve. Gases make this practice dangerous with gas-fired dryers.

Mechanical humidifiers also are available, though many efficiency experts doubt their ability to save. Rated at 200 to 500 watts, they apparently can consume as many energy dollars vaporizing the water as they can save in lowered thermostat settings.

There are two types. One attaches to the furnace and sprays vaporized water into the plenum. The other, a self-contained type, produces vapor by sending air through a moist pad.

Energy efficient pools

Even on its good days, a conventionally heated swimming pool contributes toward the total energy bill. On bad days the amount can be stunning. However, pools are used primarily in the sunny season, which means solar power can replace conventional power to a large degree.

Pool blankets, flexible covers that fit over the pool's surface, can act as passive solar devices or very effective insulators. They are a major opportunity to save energy. And nowhere else are active solar heaters so efficient as with pools.

Add to these relatively inexpensive devices a thoughtful maintenance program and a braver attitude toward water around 75°F/24°C, rather than 82°F/28°C, and the pool's contribution to the energy bill may shrink to bearable proportions.

The case for a pool blanket

A blanket decreases by 40 to 70 percent the amount of fuel needed to keep pool water at a "comfortable" range of 78° to 80°F (26° to 27°C). Blankets insulate the pool; some also allow the pool to collect heat from the sun, often eliminating the need for heating fuel in summer, and possibly extending the swimming season by a month or more.

By helping to keep debris out of the pool, a blanket also can reduce the amount of time and electricity needed to operate the pump and filter system. And, heating and filtering the pool with a blanket in place reduces the heat loss caused by water agitation.

If you're considering solar heating, a pool blanket can also help reduce the required size of an active system.

Blanket types. Basically there are four types of pool blankets: transparent, translucent, opaque foam (insulating), and opaque plastic (noninsulating).

Transparent and translucent blankets are sometimes called solar blankets, because they allow a major portion of the sunlight that hits them to penetrate deeply into the pool, warming the water. They are especially effective for unheated pools that receive unobstructed sunlight. And if you uncover your pool no more than a couple of hours a day, a solar blanket may be the most efficient type for you.

Opaque foam blankets, unless removed during the day, will greatly reduce the amount of heat gained from the sun. Also called thermal covers, they are generally made of nonabsorbent plastic foam with an upper layer of tear-resistant material. Floated on the water's surface, they are especially effective insulators at night (able to reduce a normal temperature drop of 5°F to 1°F (2.8°C to .6°C).

Opaque plastic sheets are usually made of light to dark woven materials and are anchored above the water at the sides of the pool. Sunlight does not readily penetrate such blankets, but dark or black ones will absorb enough heat to warm the top few inches of water.

If you elect to buy a blanket, look in the Yellow Pages under "Swimming Pool Enclosures" or "Swimming Pool Equipment." If you prefer to make your own, you'll find several materials in the section below.

About safety. Floating blankets will not protect children and nonswimmers who might fall

Pool blankets

Installing a pool blanket is not a complex task, whether it is of the insulating or solar type.

Many insulating blankets come as kits; some sort of fastening device secures the blanket to the edges of the pool. The most elaborate ones use rails and a powered winch to roll the blanket up. Others use simple mechanical fasteners.

Most solar blankets are just lengths of bubble plastic, the kind used as packing material. You buy it, cut strips to length, then float the film on the pool. Some pool owners frame the lengths with wood or PVC pipe, but this requires ample storage space. It is easier to roll up the plastic between uses.

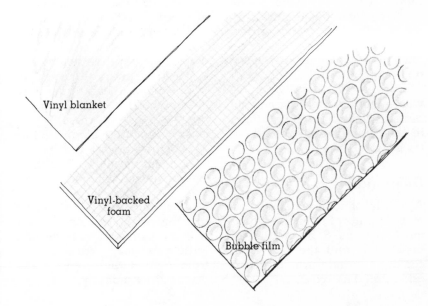

Vinyl blanket

Vinyl-backed foam

Bubble film

into the pool. (A few covers are designed to do so, but no one should be allowed to walk on any blanket.) All blankets should be removed *completely* before anyone uses the pool. It is all too easy for someone to swim or dive under a blanket, become confused, panic, and drown. Indeed it has happened too often.

The case for solar heaters

Solar heating and swimming pools are compatible for two reasons. First, the swimming season is the season of bright sun and long days, so plenty of solar heat is available when most needed. Second, heating a pool requires fewer, and less complicated components than heating a house. There are three reasons for this: pools do not need a tremendous amount of high-intensity heat; pools act as their own heat storage; and pool water can be used directly in solar collectors (can act as its own heat-transfer liquid).

Solar pool heaters are active liquid systems. The collector panels may be mounted on a south-facing roof surface (of the house, garage, or pool house),

or on an earth bank or wood framework constructed nearby to hold them.

To heat the pool, an electric pump cycles pool water through the collectors and back to the pool again. A thermostat turns on the pump when the collectors are hot enough to benefit the pool. Most systems include an auxiliary heater (usually gas fired) and a regular pool filter.

Conventional flat plate solar collectors are commercially available in a wide range of prices, output capacities, and are equally variable in durability. For the long haul, these are what you will need. But if you wish to explore the feasibility of solar heat for your pool for minimum cost, or if you simply want a short term solution, polyvinyl chloride tubing laid in parallel rows or coiled in concentric circles will operate efficiently for 5 to 10 years before beginning to break down. They can be housed in black boxes or left exposed; their temperatures never are high enough to damage roofing.

The plumbing arrangement is much the same as for solar water heaters (see page 88), but design factors are highly com-

plex, and best entrusted to a professional designer.

Pool operation

Several conservation measures can result in savings of conventional energy beyond those afforded by pool blankets.

• Lower the water temperature a few degrees. Heating the pool from 78° to 82°F (26° to 28°C) can increase energy cost as much as 40 percent. The greater the difference between air and pool temperature, the greater the cost of heating the water.

• Turn off the pilot light, as well as the heater, when you are away on vacation or when the pool is not in regular use.

• Have an annual maintenance check on the gas heater.

• Landscape to reduce wind around the pool. Fencing, shrubs, or trees that block wind also reduce heat loss caused by evaporation. As long as they don't block sun, you gain.

• Painting the pool black has limited value. Regardless of color, the pool will absorb 75 to 80 percent of the sun heat that strikes it. A black pool absorbs only 10 to 15 percent more than a white one.

Hand-crank rollers

You can build your own simple winch to roll up either insulating or solar blankets. The modest cost is more than repaid in ease of handling and storing the blanket.

The wooden type shown here uses 4 by 4 uprights anchored in concrete pads next to the pool decking, a good solution where decking is narrow or absent.

The one of aluminum and galvanized 2-inch pipes anchors in holes punched through pool decking by a masonry drill to accommodate the uprights.

The key measurement is height of the spindle above deck level; the easiest way to measure is to wind the blanket onto the spindle and add 6 inches to the radius.

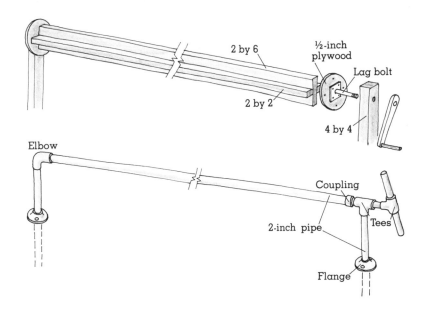

MAJOR APPLIANCES

Major appliances use a considerable part of your monthly energy supply, and, with the exception of water heaters, they tend to be relentless about it.

Substantial savings are possible in the case of water heaters partly because they consume much more energy than other appliances, partly because changing habits can reduce demand on them in major ways, and partly because their efficiency can be improved by the addition of insulation to the tank and to pipes.

Only at replacement time can you expect to save dramatic amounts of energy by installing newer model washers, dryers, dishwashers, refrigerators, freezers, and ranges, which can be far more energy efficient than older models. But there are at least a few ways to trim energy use.

A federal system of Energy Efficiency Ratios is now in effect for room air conditioners. Several other appliances are ticketed by federal law with statements of annual fuel use to aid comparison shopping. But even with these, measuring the profitability of replacing an existing appliance remains difficult.

One solution is to learn which design factors lead to energy efficiency, and to check your appliances for how well they measure up.

Savings with Major Appliances and Lighting

General energy efficiency of a house does not markedly affect performance of appliances.

Action	Range of savings	Your potential savings
Water heater Adding insulation to inefficient tank (See page 87)	If heater is in unheated space, savings can approach 10% of water heating bill; if heater is in heated room, probable savings are closer to 5%.	
Insulating hot water pipes to R-4 (See page 87)	Maximum predictable savings are 6% where pipes run in unheated basement or crawlspace.	
Installing 2-gallon-per-minute flow restrictor on showerhead	Cuts hot water use in half for each shower; probable savings are about 5% of total use.	
Adding effective solar heating system (See pages 88–90)	Most experts agree that savings range between 50 and 60% of energy used to heat water.	
Clothes washer Efficient use and maintenance (See page 92)	Maximum estimated savings: about 20%.	
Clothes dryer Efficient use and maintenance (See page 92)	Maximum estimated savings: about 15%.	
Dishwasher Efficient use and maintenance (See page 90)	Maximum estimated savings: about 15%.	
Range, oven, etc. Efficient use and maintenance (See page 93)	Maximum estimated savings: about 5%.	
Refrigerator-freezer Efficient use and maintenance (See page 91)	Estimates are bewilderingly diverse, from a high of 50% to a low of 5%.	
Lighting Replacing incandescent fixtures with fluorescent (See page 93)	Savings are 50% for each fixture replaced.	

Water heaters

The water heater for a family of four uses as much as 6,000 kwh per year. Combining an efficient heater with wise use patterns can cut this figure in half.

Design factors. In both gas and electric models, conventional storage tank heaters come in capacities ranging from 20 to 60 gallons. The old rule of thumb for efficient capacity is to allow 26 gallons for two, 40 gallons for a family of four. A new measure of capacity is the "first hour rating," the number of gallons of hot water the tank can deliver in one hour of continuous use. Look for the highest rating for the lowest annual operating cost, still with the idea that two people need 26 gallons of available hot water, and four need 40.

The key design factor in energy-efficient water heaters is two inches of foam insulation. (Many less efficient models have less than two inches of fiberglass.)

The inherent flaw in storage tank heaters is that they heat the same water twice while you're playing tennis, then heat a whole new tankful as you shower. One answer to this problem is to install a small, on-demand water heater. Long familiar to travelers in Europe, these rapidly heat as little as a gallon of water at a time for such specific tasks as shaving or hand-washing. Because they do not store hot water between uses, they are potentially efficient. The difficulty is integrating them efficiently into a system with a conventional storage tank.

Tuning up your system. As the chart on the facing page shows, tank and hot water pipe insulation are the primary ways to improve water heater performance, especially where one or both are in unheated space.

(Continued on page 90)

Extra insulation for water heaters

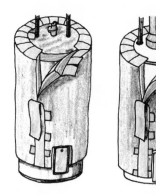

Kits usually contain R6 insulation (see R-value in Glossary, page 94) of 1½-inch-thick fiberglass or ⅜-inch-thick foam inside a vinyl or similar cover, plus tape to seal joints. Those for electric heaters cover both top and sides; those for gas heaters cover only sides. To install, all you need is a tape measure and pair of scissors; mark and cut out holes to expose controls and the drain valve.

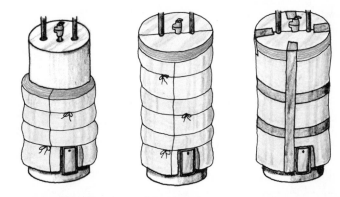

If you make a blanket for an electric heater, use R6 or R9 fiberglass batt insulation with a vapor barrier. Cut batt to lengths equal to circumference of tank; then wrap lengths around tank, beginning at bottom. Cut holes to expose controls as you go. Hold lengths temporarily with string. After cutting disc for top, tape it to top wall section, then tape all horizontal joints. Tape vertical joint last after removing strings.

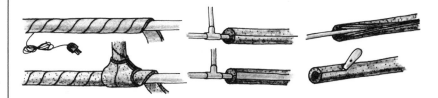

Several types of insulation are available for hot water pipes in unheated spaces. Spiral wrap tape with electric resistance wires inside may be required in severe winter climates. Spiral wrap foam tape is another possibility. Most common, though, is the foam tube, preslit or not. Some preslits have a zipper lock; most must be taped shut.

IS SOLAR WATER HEATING FOR YOU?

Of all the applications of solar power, heating water is the most practical with today's technology.

Most systems are designed to provide about 60 percent of a family's annual hot water needs—the entire supply in warm months, lesser amounts in cold and cloudy seasons.

Solar-heating choices

Solar water heaters are economical—they pay for themselves within 7 years—and can be added to almost any existing house. Of the three basic systems, "breadbox" or "batch," thermosiphon, and active, the breadbox or batch heater is most within the reach of the plumbing and carpentry skills of the average weekend builder. The other types almost always require professional installers. But even the breadbox system must be designed by an experienced solar engineer if it's to work at peak efficiency.

Breadbox or batch. This passive system relies on the pressure of incoming water rather than on pumps.

Collector and storage are the same unit—a glass-lined water heater tank stripped of its outer casing, insulation, and heating mechanism.

The tank is painted flat black to absorb solar radiation and housed in an insulated box—the breadbox—with south-facing double glazing to trap heat. The interior of the box may be painted

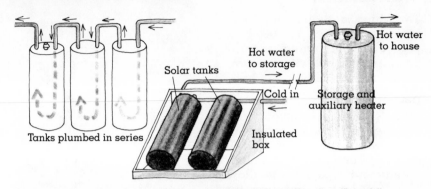

Solar tanks

Hot water to storage

Cold in

Hot water to house

Storage and auxiliary heater

Tanks plumbed in series

Insulated box

Passive solar water heating—usually called "breadbox" or "batch"—requires no pumps or sensors. Water simply gets warmer and warmer as normal pressure pushes it through a series of tanks enclosed in collector-like box. (See page 90 for construction hints.) Though its design and installation are simple, water heats slowly, tanks are very heavy, and insulating cover on box needs frequent manipulation.

flat black or may have a parabolic reflector to focus sun rays on the tank. In all but the mildest climates, the box must have an insulated lid that you close at night and on cloudy days to prevent heat loss.

Though most systems rely on one or two 40 or 60-gallon tanks to preheat water on its way to a conventional heater, some systems use as many as six tanks plumbed in series to heat the water to usable temperature during the sunny season.

In severe climates, you may want to plumb shutoff valves into the line so the collector tanks can be bypassed during prolonged unfavorable weather. No other controls are needed, since you're simply diverting household water through an extra loop on its way to your conventional water heater; pressure in the main keeps the water moving as it's needed.

There are drawbacks to this system. Since the solar tank(s) must be out-

doors, a passive system cannot store heat overnight as well as an active system; you must faithfully open and close the insulated lid as the sun comes and goes. In any weather, water heats very slowly—there may not be enough hot water for bathing until near dusk.

Because the full tanks are very heavy (water weighs 8.33 pounds per gallon), mounting them on the roof is inadvisable unless a structural engineer determines that the roof can bear the weight.

The breadbox can be installed at ground level, providing it can be oriented to the south.

Construction hints for this system are on page 90.

Thermosiphons. A thermosiphoning system provides full-scale passive water heating. In all outward respects, a thermosiphoning water heater resembles a small, active liquid space-heating system with collectors and a

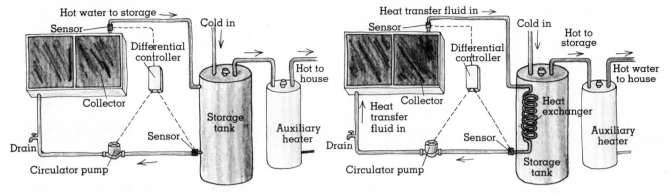

Active solar, open-loop water heating system sends regular household water through collectors on way to storage tank. Sensors and differential controller activate pump when temperature at collectors is warmer than at storage tank, stop pump when reverse is true. Back-up heater can be shut down in season and serves as year-round storage. Once popular, system now seldom used for health reasons.

Active solar, closed-loop water heating system sends liquid other than water through collectors for heating; heat exchanger inside or wrapped around storage tank transfers solar heat to household water supply in storage tank. Other elements are same as open loop. Advantages over open loop: antifreeze allows use during freezing weather; minimal risk of leaks contaminating water.

storage tank. The difference is that it has no pumps. Instead, a liquid circulates through the collectors by natural convection, rising and falling in response to the sun's heat just as air does.

To initiate this convective current, the collectors are mounted with their tops *below* the bottom of the well-insulated tank into which they feed their heated liquid.

An extra storage tank between the collectors and the conventional water heater allows water to become fully heated before going to the conventional heater for use. A heat exchanger, usually an external jacket, transfers heat from the collector loop to the water in the heavily insulated storage tank.

As in most systems using collectors, the liquid in the system is antifreeze. Collectors cannot withstand the internal pressures of forming ice; the antifreeze eliminates the

need to drain the system at every approach of frost.

When added to an existing building, a typical thermosiphoning system has its collectors at ground level and the storage tanks at some higher point in the house.

Active systems. The most familiar design of active systems places the collectors on the roof and the storage tanks inside the house, usually on the ground floor or in the basement.

With this arrangement, you'll need to use pumps, valves, and automatic temperature-sensing controls—the elements that make a system active as opposed to passive.

The diagrams above show the typical workings of both closed and open-loop active systems. In the past, the open-loop system was preferred because of its simpler plumbing and slightly greater efficiency in heating water. But leaks in the system some-

times contaminated household water. Now, for reasons of health, virtually all active systems (and other systems using collectors) are closed-loop designs.

Collectors must be oriented to within 20° of true south and tilted to within 10° of the latitude of the area where they're located.

Opinions vary on the amount of collector surface required. Some experts recommend 1 to 2 square feet of surface per gallon of water to be heated; others say as little as ½ square foot per gallon will do.

All generally agree that a system should be designed to heat 60 to 80 gallons per day for a household of four.

Both the complexity of the plumbing and the complications of routing it and the control wiring through an existing building put installation of active systems beyond the reach of do-it-yourselfers.

. . . *Continued from page 87*

"Breadbox" water heater design

"Breadbox" or "batch" water heaters demand only elementary carpentering and plumbing skills. Large units using several tanks are no more complex than single-tank installations.

See the section on pages 88–89 for general design factors.

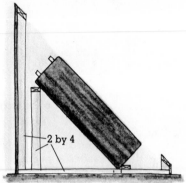

1. Using 2 by 4s, build a simple frame sized to allow water tanks to rest, preferably at an angle between 35 and 45°. There should be insulation between an inner and outer skin of the box, bottom and sides. Paint the inner skin black.

2. The tanks are placed in the box and plumbed in series before the box cover is added. Cold water line is plumbed to cold water inlet of first tank; hot water line to hot water outlet of last tank; connectors run hot to cold of intervening tanks.

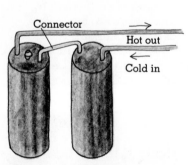

3. Once tanks are in place and plumbed, all exposed pipe should be insulated to R-4. At this time, you may also wish to plumb a common drain line to drain valves on each tank, to facilitate maintenance. In any case, provide for drainage.

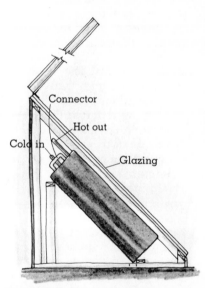

4. With tanks in place and plumbing complete, top of tank goes in place. Angled side is covered with double-glazed window, then with top-hinged insulating cover (usually exterior grade plywood outer surface, rigid board insulation inner surface).

If the wall of a water heater feels distinctly warm to the touch about halfway up, there is profit in adding a layer of insulation. The greater the difference in temperature between the hot water inside the tank and the colder air outside it, the greater the profit of insulating.

Several manufacturers offer kits for both electric and gas heaters, or you can cloak your electric heater with regular batt fiberglass insulation. This won't do for gas heaters; vapor barriers on batt insulation are flammable and the risk of fire is too great.

Maintenance. Regular draining at the bottom valve of gas heaters rids storage tanks of sediment, which insulates the water against heat if allowed to build up. Recommended intervals: at least every 6 months, at most, monthly.

Repair leaky hot water faucets (washers, washer seats, or packing).

Operating tips. Aside from setting the water heater thermostat down to pilot on gas-fired heaters, or low on electric heaters, during prolonged absence, you can install a 2-gallon-per-minute flow restrictor at the showerhead. Other than these, most tips for saving energy deal with the appliances that use hot water rather than with the heater itself.

Dishwashers

Dishwashers do not lend themselves to modification for energy efficiency. Happily, they are efficient by nature. Though they require the water heater to be set at 140°F (20° higher than needed for other appliances), most models use less hot water than would be needed to wash the same number of dishes by hand. (The celsius figures are

60°C, or 11° higher than for other appliances.)

Design factors. The machines use some power to pump wash water, and much more to dry dishes. Thus, the major factor in efficiency is an optional off switch for the drying cycle. (You prop the door partially open and let the dishes air-dry instead.) Other energy saving features are a small water capacity, and a special short cycle for small loads.

Maintenance. Except in self-cleaning models, clean the filter screen at the bottom of the washing compartment frequently.

Operating tips. The major chance to save is by air-drying the dishes instead of using the drying cycle. Others:
• Run only full loads, and use special short cycles for easily washed loads (ones with no animal fats or baked-on food particles).
• Scrape the dishes well before loading; if you rinse, use cold water.
• Load according to the manufacturer's instructions.

Clothes washers

Clothes washers are more variable in design and operation than their humble job might suggest, but the key to efficiency lies with water use rather than the machine's motor.

Design factors. High efficiency machines are those that use the fewest gallons of hot water per pound of clothes by means of variable load settings. Other energy saving features are variable water temperature settings for wash cycles and an automatic cold water rinse. (Hot water rinses do not affect cleansing quality.)

Operating tips. The key is using a minimum of hot water by:
• Filling to load level only.
• Using the coolest wash water suitable. (Some lightly soiled washes can be done entirely in cold water.)
• Rinsing in cold water.
• Pretreating grease spots or other stubborn stains.
• Presoaking heavily soiled clothes.
• Sorting fabrics by type and weight to permit minimum use of hot water.

Dryers

As with most appliances, little can be done to modify the machine you buy.

Design factors. Whether you choose a gas-fired or electric dryer depends on which energy source your local public utility has most abundantly.

Maintenance. Lint should be cleaned from the motor housing three or four times a year; see manufacturer's instructions.

Operating tips. The most effective way to conserve is to replace the dryer with a clothesline and sunshine whenever weather permits. Also:
• Have clothes as dry as possible before putting them in the dryer.
• Run consecutive loads to take advantage of accumulated heat.
• Clean the lint screen after each load.
• Sort clothes according to fabric weight; then use the lowest temperature setting possible.
• Clean the outside vent monthly.

Refrigerators and freezers

Some inefficient refrigerator-freezers can be improved by the addition of an extra layer of insulation, but for the most part energy savings result from wise purchase and thoughtful use.

Small appliances

To help you sort out your energy use, we show here the amounts of electricity consumed by a variety of small appliances. The number of uses is average for a family of four in one year.

Appliance	Typical wattage	Number of uses	kwh
Blanket	150	180	108.0
Heating pad	60	52	3.4
Hair dryer			
—soft bonnet	400	100	30.0
—hard bonnet	900	100	45.0
—hand-held	1000	250	33.3
Shaver	15	365	0.5
Iron	1100	52	59.5
Coffee maker			
—percolator	600	600	138.0
—drip	1100	600	146.5
Deep fryer	1500	50	31.2
Broiler	1200	100	90.0
Frying pan	1200	180	100.4
Rotisserie	1400	26	73.0
Toaster	1100	700	38.5

Boxing in your freezer's air

Just as water heaters in unheated space waste much of their warmth, so freezers in warm kitchens waste much of their chill. Fortunately, many of the least efficient types (for example, the side-by-side refrigerator-freezers) have flat sides and flat doors that can be covered on the outside with insulating paneling.

If the appliance can be recessed into a tailored space, only the doors need be paneled. Otherwise the top and both sides need to be covered for any real gain. Covering or crowding the back of the unit interferes with the venting of warmed air, though; be sure it has a clear escape path.

Locate angle clips on refrigerator door

Then secure clips to paneling

Or...

Paneling covers furring strips

If you opt for heavy-duty insulation, sandwich a layer of aluminum foil between a thin insulation panel and your surface paneling. For lighter duty, secure furring strips to angle clips and let airspace do some of work; again, cover back of paneling with reflective foil. Remove brand plates and door handles for flat surface; replace with wooden handles.

Design factors. The common characteristics underlying high efficiency for all refrigerators, refrigerator-freezers, and freezers are heavy insulation and manual defrost.

The most efficient freezers are chest types, because cold air sinks to the bottom rather than slumping out the door.

The following results of a utility company study may be used as rough indications of energy consumption by units two years old or older:

Single door, manual-defrost refrigerators of 10 to 12 cubic feet use 480 to 600 kwh per year.

Refrigerator-freezers of 12 to 15 cubic feet, with manual-defrost top freezer and automatic-defrost refrigerator, use from 650 to 1,050 kwh.

Refrigerator-freezers of 17 to 22 cubic feet (top freezer or side by side) with fully automatic defrost use 1,100 to 1,750 kwh.

Upright freezers of 15 to 21 cubic feet, with manual defrost, use 760 to 1,000 kwh; similar units with automatic defrost use 1,020 to 1,524 kwh each year.

Manual-defrost chest freezers of 15 to 22 cubic feet use 800 to 985 kwh; 25-footers use about 1,015.

Maintenance. Brush the compressor coils and motor housing clean of lint three or four times per year.

Check the seal on door gaskets regularly. An easy check is to close the door on a dollar bill; if the bill slips easily at any point, replace the gasket.

Operating tips. The simplest and most effective step is to open the door as seldom as possible and to keep it open for the shortest time possible, especially in hot weather. Also:

• For each cubic foot of freezer space add no more than 3 pounds of food at one time, none of it warm or hot.

• Allow for air circulation around the unit so that generated heat can escape.

• Locate units away from sources of heat, especially ranges, dishwashers, direct sunlight, and heaters.

• In manual or semiautomatic-defrost models, defrost when the rime (ice crust) becomes ¼ inch thick. (Thicker rime insulates *against* the chilling ability of the compressor.)

• If your refrigerator's anti-sweat device has a switch, turn it on only when the refrigerator shows beaded moisture.

Ranges and ovens

The contemporary American cook has a remarkable number of appliances to choose from, some energy efficient and some less so. Nearly all potential savings lie with careful selection and use of what is available. The principal exception is the gas range, which can be made more efficient by the replacement of its pilot light with electronic igniters.

The major types of cooking units are electric or gas ranges, separate cooktops using either energy source, separate ovens using either energy source, convection ovens, microwave ovens, crock pots, and electric frying pans.

Design factors. Gas ranges should have electronic igniters rather than pilot lights.

All ovens should have windows—opening the door to peek drops the internal temperature about 25°F/14°C, thus prolonging cooking time.

Automatic-cleaning ovens are especially efficient. Their extra-hot cleaning cycle requires extra-thick insulation, so they hold heat better. (Continuous-cleaning ovens are not so efficient. They have only conventional insulation, because their special interior finish breaks down matter at lower temperatures.)

Both convection and microwave ovens can be more efficient than conventional ovens.

One study suggests that convection ovens (which employ a fan to distribute heat evenly) cut energy use about 30 percent compared with conventional ones. A similar study shows microwave ovens (which heat with a different length of wave) to save about 14 percent as measured against a regular oven, especially when cooking small amounts of food.

Operating tips. Pots and pans and the way you use them are the keys to an efficient cooktop. Place them on the burner before it is turned on; use pans with these features:

• A flat bottom
• Excellent conductivity (made of copper, copper-clad stainless steel, or aluminum)
• A larger diameter than the burner
• A tight-fitting lid

Conventional ovens are more efficient when:

• The door seal is kept in good repair
• There is no peeking
• More than one dish cooks at a time. Let a microwave oven replace a conventional oven for single dishes, especially when portions are small
• You close the oven door while you baste a dish
• You line the oven with foil only if the manufacturer specifies that it may be used
• A self-cleaning oven is run through a cleaning cycle immediately after a cooking cycle so that the minimum increase in heat is required for cleaning.

Lighting

Though the subject of lighting is vast, lights take only a modest part of an average energy budget and offer only modest opportunities to save. The mea-surable saving is through replacement of incandescent lights with fluorescents.

Fluorescent lights generate the same level of illumination as incandescent lights but use one-fourth to one-third the amount of energy. They also last 10 to 12 times as long as incandescents.

Kitchens, bathrooms, utility rooms, and shop or other hobby areas long have been lighted with tube fluorescents, because high levels of illumination are more important in these areas than in rooms used mainly for social purposes. Circular fluorescent lights can be used in place of tubes if esthetics or space require smaller fixtures.

Changing over from incandescent to fluorescent lighting does not produce a quick payback because of the much greater initial cost of fluorescent fixtures.

Operating tips. Where using incandescent light, choose fixtures that use one large rather than several small bulbs. One 100-watt bulb produces the same level of light as three 40-watt bulbs.

Use translucent rather than opaque shades.

Keep lights and fixtures clean. As much as 20 percent of the light generated can be lost to hazing dusts.

Take advantage of reflected light; keep portable fixtures as close as possible to light-colored walls or other surfaces. (Light diminishes by the square of the distance it travels.)

Install solid state dimmer switches or three-way bulbs where possible; use lower settings to minimize energy consumption.

Use standard rather than long-life incandescent bulbs in all but places where changing bulbs is difficult. Standard bulbs use about 20 percent less energy to achieve the same level of illumination.

Glossary of terms

Ambient. Outside air, usually defined by its temperature or relative humidity.

Active solar system. A heating or cooling system using mechanical methods to distribute collected heat.

British thermal unit. A basic measure of heat. Precisely, the amount of heat required to raise the temperature of one pound of water by one degree from 39.2°F; approximately, the amount of heat given off by burning one wood kitchen match.

Conduction. The movement of heat through matter (usually solid) without movement of the matter itself.

Convection. The transfer of heat in a liquid or gas by currents set up between unequal temperatures. (A specific example is warm air rising and cool air sinking.)

Degree days—heating. Equal to number of degrees by which the daily average temperature exceeds 65°F. Seasonal total is used to measure average annual heating requirements, insulation levels, and kindred matters.

Electronic igniter. Device which uses electric resistance to ignite gas-fired appliances in place of pilot lights.

Energy. The capability of doing work. There are several forms; in this book the term is used almost exclusively to describe thermal energy. (Other forms are kinetic, electromagnetic, and potential.)

Energy Efficiency Ratio (EER). A measure of energy consumed versus work achieved. Used to judge relative performances of similar appliances, esp. air conditioners.

Eutectic salts. Salts that melt readily at low temperatures (as low as 90°F/32°C), and, in so doing, store large quantities of latent heat which they release when cooling and resolidifying.

Greenhouse effect. Ability of glass or clear plastic to transmit short-wave solar radiation into a room or collector, but to trap long-wave heat emitted by room or collector interior; thus, the basis of solar heating systems.

Greenhouse, solar. A greenhouse designed to maximize the greenhouse effect with double glazing and efficient thermal mass.

Useful addresses

**REGIONAL ENERGY OFFICES
U.S. DEPARTMENT OF ENERGY
(USDOE)**

REGION I
Conn., Maine, Mass., N.H., R.I., VT.
USDOE Region I
150 Causeway St.
Boston, MA 02114
(617) 223-5207

REGION II
N.J., N.Y.
USDOE Region II
26 Federal Plaza
New York, NY 10278
(212) 264-4780

REGION III
D.C., Del., Md., Pa., Va., W.V.
USDOE Region III
1421 Cherry St.
Philadelphia, PA 19102
(215) 597-4319

REGION IV
Ala., Fla., Ga., Ky., Miss., N.C., S.C.
USDOE Region IV
1655 Peachtree Street, N.E.
Atlanta, GA 30367
(404) 881-4463

REGION V
Ill., Ind., Mich., Minn., Ohio, Wis.
USDOE Region V
175 West Jackson Blvd.
Chicago, IL 60604
(312) 353-8420

REGION VI
Ark., La., N.M., Okla., Tex.
USDOE Region VI
Box 35228
2626 West Mockingbird Lane
Dallas, TX 75235
(214) 767-7736

REGION VII
Iowa, Kans., Mo., Nebr.
USDOE Region VII
324 East 11th St.
Kansas City, MO 64106
(816) 374-2941

REGION VIII
Colo., Mont., N.D., S.D., Utah, Wyo.
USDOE Region VIII
Box 26247
Lakewood, CO 80226
(303) 234-2420

REGION IX
Ariz., Calif., Hawaii, Nev.
USDOE Region IX
333 Market St.
San Francisco, CA 94105
(415) 764-7035

REGION X
Ark., Idaho, Oreg., Wash.
USDOE Region X
915 Second Ave.
Seattle, WA 98174
(206) 442-7285

STATE ENERGY OFFICES

ALABAMA
Department of Energy
3734 Atlanta Highway
Montgomery, AL 36130
(205) 832-5010

ALASKA
Energy & Power Development
338 Denali St., 7th Floor
Anchorage, AK 99501
(907) 276-0508

ARIZONA
State Energy Office
1700 West Washington St.
Phoenix, AZ 85007
(602) 255-4955

ARKANSAS
Department of Energy
3000 Kavanaugh Blvd.
Little Rock, AR 72205
(501) 371-1370

CALIFORNIA
California Energy Commission
1111 Howe Ave.
Sacramento, CA 95825
(916) 920-6103

COLORADO
Energy Conservation Office
1525 Sherman St.
Denver, CO 80203
(303) 839-2507

CONNECTICUT
Energy Division
Office of Policy & Management
80 Washington St.
Hartford, CT 06115
(203) 566-2800

DELAWARE
State Energy Office
Box 1401
56 The Green
Dover, DE 19901
(302) 736-5644

FLORIDA
Governor's Energy Office
301 Bryant Bldg.
Tallahassee, FL 32301
(904) 488-6764

GEORGIA
Office of Energy Resources
270 Washington St.
Atlanta, GA 30334
(404) 656-3874

HAWAII
Planning & Economic
  Development
Box 2359
Honolulu, HI 96804
(808) 548-3033

IDAHO
Office of Energy
State House
Boise, ID 83720
(208) 334-3800

ILLINOIS
Institute of Natural Resources
325 West Adams
Springfield, IL 62706
Fuel Allocation Hotline:
  (217) 782-1986
Director's Office: (217) 785-2002

INDIANA
State Energy Office
440 North Meridian St.
Indianapolis, IN 46204
(317) 232-8940

IOWA
Energy Policy Council
Capitol Complex
Des Moines, IA 50319
(515) 281-4420

KANSAS
State Energy Office
214 W. 6th St.
Topeka, KS 66603
(913) 296-2496

KENTUCKY
Department of Energy
Box 11888
Lexington, KY 40578
(606) 252-5535

LOUISIANA
Research & Development
Dept. of Natural Resources
Box 44156
Baton Rouge, LA 70804
(504) 342-4594

MAINE
Office of Energy Resources
Statehouse Station 53
Old Federal Bldg.
Augusta, ME 04333
(207) 289-3811

Heat exchanger. Device consisting of a long coil of metal pipe or a multifinned radiator, used to transfer heat from one fluid inside it to another outside it without bringing the two fluids into direct contact.

Fuel. Any substance that can be burned to produce heat.

Humidity, relative. Measure of the amount of moisture in air relative to the capacity of that air to maintain moisture in suspension. Expressed as per cent.

Insolation. Heat gained from the rays of the sun on any surface.

Insulation. Commonly, material resistant to heat transfer especially designed for use in construction.

Kilowatt hour. A measure of electricity: 1,000 watts transmitted for one hour. Has the potential of 3,413 BTUs.

Passive solar. A solar heating and/or cooling system using natural means of heat distribution. Principal elements usually are windows oriented to allow radiation to strike a thermal mass.

R-factor. A measure of the ability of a material to insulate against heat transfer; 1 R equals the resistance of heat transfer of wood 1 inch thick.

Radiation. The transfer of heat from a source to a solid by light waves, esp. infra-red.

Reflectance value. The ability of a material to reflect potential heat from solar radiation; usually measured as a per cent.

Therm. A measure of natural gas equal to 100 cubic feet; has 100,000 BTUs of potential heat.

Thermal mass. Material with capacity to store heat radiated from a source.

Vapor barrier. Material impermeable by moisture. In construction, often bonded to insulation to improve its efficiency.

Zone heating. The idea of using central heating only in selected areas, or of substituting localized for central heating.

MARYLAND
State Energy Office
301 West Preston St.
Baltimore, MD 21201
(301) 383-6810

MASSACHUSETTS
Office of Energy Resources
73 Tremont St.
Boston, MA 02108
(617) 727-4732

MICHIGAN
Energy Administration
Commerce Dept.
Box 30228
Lansing, MI 48909
(517) 374-9090

MINNESOTA
State Energy Agency
150 East Kellogg Blvd.
St. Paul, MN 55101
(612) 296-5120

MISSISSIPPI
Dept of Energy & Transportation
510 George St.
Jackson, MS 39202
(601) 961-5099

MISSOURI
Div. of Energy Programs
Dept. of Natural Resources
Box 176
Jefferson City, MO 65102
(314) 751-4000

MONTANA
Dept. of Natural Resources &
    Conservation
32 South Ewing
Helena, MT 59601
(406) 449-3647

NEBRASKA
State Energy Office
State Capitol Bldg.
Box 95085
Lincoln, NE 68509
(402) 471-2867

NEVADA
Dept. of Energy
400 W. King St.
Carson City, NV 89710
(702) 885-5157

NEW HAMPSHIRE
Governor's Council on Energy
2½ Beacon St.
Concord, NH 03301
(603) 271-2711

NEW JERSEY
State Energy Office
101 Commerce St.
Newark, NJ 07102
(201) 648-2744

NEW MEXICO
Energy & Minerals Dept.
Box 2770
Santa Fe, NM 87501
(505) 827-2471

NEW YORK
State Energy Office
2 Rockefeller Plaza
Albany, NY 12223
(518) 473-4376

NORTH CAROLINA
State Energy Division
Box 25249
430 North Salisbury
Raleigh, NC 27611
(919) 733-2230

NORTH DAKOTA
State Energy Office
1533 North 12th St.
Bismarck, ND 58501
(701) 224-2250

OHIO
State Energy Dept.
30 East Broad St.
Columbus, OH 43215
(614) 466-6797

OKLAHOMA
State Energy Dept.
4400 N. Lincoln Blvd.
Oklahoma City, OK 73105
(405) 521-3941/2995

OREGON
Department of Energy
102 Labor & Industries Bldg.
Salem, OR 97301
(503) 378-4040

PENNSYLVANIA
Governor's Energy Council
1625 North Front St.
Harrisburg, PA 17102
(717) 783-8610

RHODE ISLAND
Governor's Energy Office
80 Dean St.
Providence, RI 02903
(401) 277-3370

SOUTH CAROLINA
Division of Energy Resources
1122 Lady Street
Columbia, SC 29201
(803) 758-2050

SOUTH DAKOTA
Office of Energy Policy
Capitol Lake Plaza
Pierre, SD 57501
(605) 773-3603

TENNESSEE
State Energy Authority
226 Capitol Blvd.
Nashville, TN 37219
(615) 741-6671

TEXAS
Energy Advisory Council
200 E. 18th St.
Austin, TX 78701
(512) 475-0414

UTAH
State Energy Office
825 N. 3rd West
Salt Lake City, UT 84103
(801) 533-5424

VERMONT
State Energy Office
State Office Building
Montpelier, VT 05602
(802) 828-2393

VIRGINIA
Energy & Emergency Services
310 Turner Road
Richmond, VA 23225
(804) 745-3245

WASHINGTON
State Energy Office
400 East Union St.
Olympia, WA 98504
(206) 754-0700

WEST VIRGINIA
State Fuel & Energy Office
1591 Washington St. East
Charlestown, WV 25311
(304) 348-8860

WISCONSIN
Division of State Energy
101 S. Webster St.
Madison, WI 53702
(608) 266-8234

WYOMING
Energy Conservation Office
320 West 25th St.
Cheyenne, WY 82002
(307) 777-7131

INDEX